WI

KINDRED BEINGS

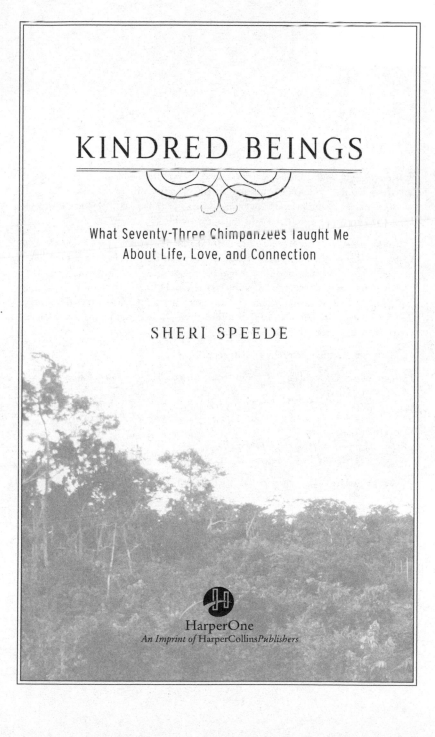

KINDRED BEINGS

What Seventy-Three Chimpanzees Taught Me
About Life, Love, and Connection

SHERI SPEEDE

HarperOne
An Imprint of HarperCollins*Publishers*

This is a work of nonfiction. The events and experiences detailed herein are all true and have been faithfully rendered as the author remembered them, to the best of her ability. Some names, identities, and circumstances have been changed to protect the privacy and/or anonymity of the various individuals involved.

KINDRED BEINGS: *What Seventy-Three Chimpanzees Taught Me About Life, Love, and Connection.* Copyright © 2013 by Sheri Speede. All rights reserved. Printed in the United States of America. No part of this book may be used or reproduced in any manner whatsoever without written permission except in the case of brief quotations embodied in critical articles and reviews. HarperCollins books may be purchased for educational, business, or sales promotional use. For information please e-mail the Special Markets Department at SPsales@harpercollins.com.

HarperCollins website: http://www.harpercollins.com

HarperCollins®, 📖®, and HarperOne™ are trademarks of HarperCollins Publishers.

FIRST EDITION

Designed by Claudia Smelser
Photograph on title page courtesy of the author
The credits on pages 255 and 256 constitute a continuation of this copyright page.

Library of Congress Cataloging-in-Publication Data

Speede, Sheri.
  Kindred beings : what seventy-three chimpanzees taught me about life, love, and connection / by Sheri Speede. — First edition.
      pages cm
   ISBN 978-0-06-213248-2
   1. Chimpanzees—Behavior. 2. Cognition in animals. I. Title.
   QL737.P96S635 2013
   599.88515—dc23   2013005848

13 14 15 16 17 RRD(H) 10 9 8 7 6 5 4 3 2 1

This book is written in loving memory of
Sara Katherine and Lena Pearl.

And it is dedicated to the volunteers: You brought your open hearts and enthusiasm from Australia, Brazil, Cameroon, Canada, England, Finland, France, Holland, Hungary, Ireland, Israel, New Zealand, Norway, Poland, Portugal, South Africa, Spain, Sweden, Switzerland, and the United States. You survived pit latrines, cold baths, couscous, cockroaches, long exhausting workdays with far too little rest, and candlelit nights for months on end. Sanaga-Yong Chimpanzee Rescue Center was built on your backs. Only a few of you are mentioned by name in the following pages, but with deep gratitude and respect, I dedicate this book to all of you.

CONTENTS

CONTENTS

INTRODUCTION

On September 24, 2008, beloved elder chimpanzee Dorothy lay down on the grass at the edge of the forest in a somewhat obscure African sanctuary and died. About five decades earlier, when Dorothy was an infant, poachers supplying the illegal ape meat trade killed her mother and took her captive. She spent most of her sad life chained by her neck as a hotel tourist attraction, but she died among friends who loved her at Sanaga-Yong Chimpanzee Rescue Center in Cameroon's Mbargue Forest.

The morning after Dorothy's death we conducted a small funeral service for volunteers, our African staff, and people from the village community who came to pay their respects. Afterward, Dorothy's longtime caregiver, Assou Francois, pushed her body in a creaky wheelbarrow toward her gravesite, which had been prepared beside the twenty-acre forested enclosure where she had lived. With a small procession of staff and volunteers, I followed behind. As we neared the enclosure, the twenty-five chimpanzees who had lived with Dorothy heard the wheelbarrow and came out of the forest. As they lined up at the fence line, straining to see her body, I instructed Assou to pull the wheelbarrow close to the fence and stop. As I caressed Dorothy's head, and the chimpanzees she loved best gazed at her a final time in silent grief, volunteer Monica Szczupider snapped a photo.

After we buried Dorothy, I saw Monica's picture and hardly gave it a second thought, but this snapshot of emotion soon would be seen around the world. After Monica won a *National Geographic* photo contest and the magazine published the funeral photo in a glossy double-page spread, numerous other magazines and newspapers also published it. Several journalists interviewed me about it. Invariably, they asked me if I had been surprised by the chimpanzees' reactions to Dorothy's death.

"No, I wasn't surprised in the slightest," I always answered honestly.

After working closely with chimpanzees for years, I took for granted their capacity for a broad range of deep emotions. I had always been deeply sympathetic to the suffering of animals; their particular vulnerability and innocence awakened the compassionate defender in me, enough so that I had dedicated my career to it even before coming to Africa. But my direct experience with captive adult chimpanzees was something different. They were so much more similar to me than either of us was to any other animal. In these chimpanzees I recognized another kind of people, like me in many ways, unlike me in others. They were also animals, they were also apes, and so was I an animal and an ape. In the face of the chimpanzees' profoundly familiar ape consciousness and in the genuine friendships that grew between us, I became a more fully realized human animal. I knew chimpanzees to be charismatic and complicated. Not all were always nice. I had seen callous cruelty in their hierarchical societies, and I also had seen kindness and compassion. As the founder and director of this African sanctuary, I was equally committed to every single chimpanzee who lived here, but I cannot say I liked them all equally. Dorothy was kind. I admired her, I loved her, and I knew the chimpanzees loved her too.

Because I knew Dorothy and for years had observed her role in her chimpanzee society, I wasn't surprised by the chimpanzees' grief over her death. The human reaction to Monica's photo was a different matter; it *did* surprise me. Although we share more than 98 percent of our DNA with chimpanzees, and this genetic similarity had become common knowledge, often cited by popular media, I knew that few human people could really comprehend the intelligence and emotional complexity of chimpanzees any more than I had understood it before I worked with them. That this photo showing a simple expression of grief drew such intense interest around the world told me that many of *my kind* might have opened their hearts to a real understanding that among us animals there is an evolutionary continuum. My initial inspiration to write this book sprang from the world's reaction to the photo of Dorothy's funeral procession. My memory of her life was a compelling inspiration throughout it.

Chimpanzees are still killed for meat, taken captive as pets, and cruelly exploited in biomedical and entertainment industries. The stories of Dorothy and her circle of friends and family need to be told and understood. I tell the stories as honestly as I can, not as an unbiased scientist, but more as a loving ambassador who has attempted to understand them. My personal story, while certainly not as important, is inextricably linked to theirs.

one

ROAD TO CAMEROON

I was born to a blue-collar family in Jackson, Mississippi, the heart of social conservatism, racial segregation as a matter of right and wrong, and the Baptist Bible Belt. My mother was a very smart and loving woman who appreciated beauty in nature as much as anyone I've ever known. She had a great sense of humor, but also an underlying sadness that affected her, and her family, throughout her life. Perhaps it was her suffering that also gave rise to her sweet sensitivity for the world's vulnerable and downtrodden, which seemed out of place in the 1960s and 1970s South. My father was a firefighter who anticipated every hunting season with the excitement of a child's countdown to Christmas. Although he brought home the meat of the wild animals he killed many times, I remember him bringing home a whole deer to clean in our backyard only once, when I was quite young. Standing at our back door in my pajamas one winter evening, I watched Daddy, blue eyes twinkling, proud and triumphant, standing over the body of that beautiful buck as he lifted the head by a long antler to facilitate my full appreciation. I took one look at the pretty face and the glazed, lifeless brown eyes and ran to my room sobbing. I would never be a hunter, and as it turned out, my younger brother never took to it very enthusiastically either. I suppose we were a disappointment, but my father tried to make the best of it. He took my mother, my younger brother, and me camping every summer on the Pearl River's sandbars, where we swam, water-skied, built bonfires, and fished. For the sake of parental tolerance and my love of fried fish and hush puppies, I managed to mostly sublimate my tender feelings for fish. We went weeks without baths or telephones while my father's beard seemed to grow longer with every Miller Lite. This was his element, where I thought he was the most competent person in the world. The older I got, the less I liked these escapes from modernity, but they taught me skills and a tolerance for uncomfortable living conditions that would one day be valuable in my travels through rural Africa. Nothing about the outdoors frightened me.

I went to college at Louisiana State University (LSU) in Baton Rouge, three hours from my Mississippi home. During my freshman year on campus, I met an instructor of a Korean martial art called tae kwon do and soon became his student at the school he operated with a partner. Training at the school of these skilled and dedicated teachers awakened the athlete in me. I took my training seriously, and while I was still in college I earned my first-degree black belt; years later I earned my second-degree. Tae kwon do gave me something more than physical fitness and self-defense skills. It taught me to act in the face of fear, including my fear of failure.

Throughout my undergraduate years at LSU, I worked as a waitress to help pay my bills. For a while I made decent tips in country-and-western bars during those urban cowboy days when mechanical bulls became a part of nightlife culture in Louisiana. My parents were delighted when I was accepted into the LSU School of Veterinary Medicine, but it would mean four years of economic hardship for them, as I wasn't able to work much in the face of the rigorous academic program. They struggled to see me through it.

During my first year of veterinary school, I witnessed my "mentors" cruelly dehorning cows and castrating baby pigs, both without anesthesia, as was standard practice in the agriculture industry. Soon afterward, I swore off eating mammals forever. I wouldn't play a part in that kind of animal suffering, even as a consumer. A few years later I decided that chickens and fish were safe from me, too, thereby coming to my vegetarianism gradually.

As a young veterinarian I was restless for change and adventure, enough so that I was willing to move with my dogs and cats and few possessions to new places and endure the loneliness of being a stranger to everyone. I'm not sure what I was looking for exactly, maybe just a place where I could really feel at home. I moved from Louisiana to my first job just outside of Nashville, Tennessee, then to a clinic in Santa Fe, New Mexico, and finally landed in a bigger veterinary hospital in Portland, Oregon, where I settled down, or so I thought. I worked very hard proving my worth, and after only a year, the two owners of Pacific Veterinary Hospital offered to finance my buy-in to the practice. I became a one-third owner, making my parents very proud. For a few years I was content saving animal lives and interacting with the humans who loved them. The people who stood before me seeking help for their beloved pets were motivated, at least during the time they were with me, by what was best in them—their love and concern for someone vulnerable who depended on them, someone they *chose* to care about. After a few years, though, I got thirsty for challenge, for some adrenaline in my life, and I just *felt* I should be doing something different. Against the advice of my parents and other cooler heads, I sold my interest in the veterinary practice to take a lower-paying job with In Defense of Animals (IDA), a nonprofit animal advocacy organization based in Mill Valley, California. I opened its new Northwest office in Portland and became the first Northwest director.

"I want to use my credibility as a veterinarian to advocate for animals in a bigger way," I said to anyone who asked. Making a decent living—maximizing my earning potential—seemed by comparison a trivial goal.

During my first three years with IDA, I benefited from the mentorship and friendship of its founder and director, Dr. Elliot Katz, who was also a veterinarian. I called him frequently for advice, and he was usually a willing sounding board for my ideas. On behalf of the organization, I filed two successful public records lawsuits to get information about biomedical experiments on monkeys, and I led an effort that ended the sale of dogs and cats from an Oregon animal shelter for biomedical research. In addition and perhaps most consequentially in the long run, as a representative of IDA I was able to provide veterinary care to animals in sanctuaries, eventually including primates in Cameroon, Africa.

I met Peter Jenkins and Liza Gadsby, native Oregonians who had cofounded the conservation organization Pandrillus for drill monkeys and chimpanzees in Nigeria and had recently begun a partnership with the government in neighboring Cameroon to convert the dilapidated Limbe Zoo into the Limbe Wildlife Center (LWC). This zoo-turned-wildlife-center, in the pretty but impoverished coastal town of Limbe, served to educate visitors about the endangered status of wildlife while it acted as a primate sanctuary, receiving and caring for orphaned chimpanzees, gorillas, and monkeys. Stretched thin by their bigger project in Nigeria, Peter and Liza needed veterinary supplies and skill in Cameroon, and I was eager to help them. The exoticism of the location and the opportunity to work with chimpanzees and gorillas held a huge attraction for me.

In January 1997, sponsored in part by Dr. Shirley McGreal and her renowned International Primate Protection League and accompanied by Kathy Pearson, a technician from my old veterinary clinic, I spent one month in Cameroon providing veterinary care at LWC, mostly to young chimpanzees, gorillas, and monkeys who had been orphaned by hunters who'd killed their mothers for the meat trade.

Kathy and I landed at the airport in the city of Douala late one afternoon. We descended from the plane along a steep portable metal staircase at the edge of the runway and walked across the tarmac to the interior of the airport. The sun was bright and the air humid, hot, and familiar. While the ambient temperature and air quality were dramatically different from those of Oregon in January, it felt like going home to Mississippi. A long-term LWC volunteer named Lolly, a native of Britain in her late twenties, met us inside the airport. As we were all leaving the baggage area, hordes of aggressive porters bore down on us from every direction, speaking at us in French, all at the same time. When one tried to pry the handle of my roll-on suitcase out of my hand, we engaged in a brief but intense tug-of-war until Lolly ended it by pointing to two other men in the crowd of needy faces. They were the ones we would hire, and it was fair play I supposed, because all the men who weren't chosen fell silent and unceremoniously dispersed to find work elsewhere. When we reached the old Land Rover that Lolly would drive to Limbe, there was a dispute about the amount she would pay the porters. I couldn't follow it, but there was boisterous discussion before it was finally resolved.

As we rolled along at a snail's pace through the crowded rush-hour streets of Douala, the strange and lively scenes outside our windows absorbed Kathy and me. I was glued to the front-seat window and she to the back, hardly speaking during the trip. Vehicles on the roadway included a few packed minibuses, which I later learned to call bush taxis, a few private cars like our Land Rover, and some motorcycles, but mostly small yellow car taxis filled the road as far as I could see, more Toyotas than anything else. They all pushed forward competitively and managed to create four packed lanes on a roadway built for two. The beeping of their horns created a continual din that puzzled me at first, until I realized that the horns were positioning signals, a necessary part of the driving dance. It was how they managed to keep from hitting each other more often than they did, but judging from the scratches on the sides of all the cars, they hit each other plenty. In addition to the countless crawling cars, pedestrians also claimed the streets. Streams of dark-skinned African people, moving faster than the cars, traveled along the sides of the roadway or squeezed between the packed cars to cross it. A little boy and girl, neither older than ten, held their palms out toward us as they crossed in front of us and navigated the maze of cars to arrive on the other side of the road. *How could they be on the streets alone?* I wondered silently. Set back a yard or two from the road at various places, women sat on low stools grilling some kind of food on small barbecue pits, and people stopped to buy it. A waitress served huge bottles of beer to customers sitting at open-air tables in front of a wooden building with cracking white paint. From just inside the doors of this bar, large speakers blasted customers and commuters alike with extremely loud African music. A minute later, Dolly Parton took over the speakers with an equally loud and somewhat crackly rendition of "I Will Always Love You." A variety of pungent smells came and went along our route, creeping into the invisible fog of car exhaust. Only a few yards from the road, I saw a huge undulating pile of stinking garbage, several yards wide and five feet high in places. A number of thin dogs scrounged around its periphery.

After we broke free from the traffic jam near the outskirts of Douala, military police stopped our Land Rover to ask for Lolly's driving permit, the papers for the car, and each of our passports. Lolly stared silently ahead as she handed documents out the window. I would learn that checkpoints such as this one dot the roadways of Cameroon, manned by officers ostensibly look-ing for bandits, and certainly looking for bribes. But on this first night I experienced no sense of the intrusion, and when one grim-faced officer looked from my passport photo to my face, I smiled broadly and gave him a little wave. Caught off guard, he smiled back as he returned my passport. And off we went careening down the darkening road toward Limbe.

Kathy and I stayed in the Miramar Hotel, located in Limbe's beautiful botanical garden on a low cliff right above the rocky Atlantic coast. Our hotel room was in a row of picturesque royal blue and white cottages, nestled upon a bed of lush green mani-cured grass, accented here and there with bright tropical flowers. It had two single beds and an oscillating floor fan. On the bath-room wall hung a hot water heater, but it didn't work, and the clerk told us that at the moment none of the other rooms had hot water either. Taking cold showers was the one hardship I couldn't bear without complaint. Each morning we started our day with coffee and bread in the hotel's open-air restaurant, facing the ocean. The wood-paneled restaurant had walls on only two sides, so we had a beautiful panoramic view of the blue-gray sea, the birds gliding gracefully above it and the few fishermen moving quietly in small wooden boats on its surface.

At LWC, Kathy and I were assisted by a competent African staff who knew how to sedate primates efficiently, so we were able to accomplish a lot quickly. During our month in Limbe, we worked with the African employees and European volunteers to perform health screens on sixty-six primate orphans. Fortunately, Limbe is located in the 20 percent of Cameroon that is English speaking, so we could communicate easily here. We tested for tuberculosis and a battery of viruses, and I sutured a few wounds along the way. During our third week, we took one day off to drive a few miles out of town to the sandy beach with a South African man we met at the hotel. I ran across the dark volcanic sand like a wild thing set free and splashed into the warm Atlantic sea in shorts and a T-shirt.

Two days later, I traveled with a driver to the city of Yaoundé, six hours away, to pick up a six-year-old chimpanzee named Pierre from a French biomedical research facility that had agreed to release him to LWC. During this arduous round trip, we had to traverse Douala en route to Yaoundé. I became much more familiar with the omnipresent checkpoints and really noticed the oppressive poverty of so many of Cameroon's people. Severely disabled beggars lined the medians, and school-age children moved from car to car, either begging or selling hard candy, boiled eggs, packages of tissue, and other things. Men trying to eke out a living selling sunglasses displayed on placards hanging from their necks, or music CDs, or ties, or T-shirts, or windshield wiper blades, or any number of other things that travelers might need approached me hopefully at the service stations where we bought fuel. A dozen women and men sat under the sun in front of typewriters along a busy business route in Yaoundé, offering cheap, on-the-spot secretarial services with minimal overhead costs. I was touched by the entrepreneurial spirit of these people trying to survive in the city however they could. There were also signs of affluence in both big cities. Tiny stick-and-mud shacks abutted sprawling spacious residences and businesses.

At the biomedical research facility, we found Pierre in a small outdoor cage near the back door leading to a parking lot, and I had to move him from there to our even smaller transport cage. I had brought a blowpipe and darts with me so I could blow a dart of anesthesia into Pierre, but in the end I didn't need it. I was able to slip my hand through the bars and inject the drug into Pierre's leg with a syringe while a technician at the research facility distracted him.

I sat beside the transport cage in the back of the Land Rover so I was there to comfort Pierre when he woke up only a few minutes outside of Yaoundé. He knew that he had been stuck and I was the one who'd done it, but he accepted the bananas and papaya and peanuts I offered him through the bars of the cage. His grateful grunts of delight over the food, interspersed with his deep, probing gazes into my eyes, told me he was open to friendship, even as his eyes disconcerted me slightly with the obvious intelligence they windowed. His protruding ribs spoke volumes about the callous disregard of the research facility staff, including affluent expats from Europe, and I tried not to let my anger ruin my enjoyment of this sweet time. I hadn't yet learned to mimic the way chimpanzees groom one another, but as I touched and petted Pierre, he leaned toward me against the bars of the cage to maximize my access. When I was tired, we leaned against opposite sides of the same bars so our bodies were touching. During a stop for fuel, I asked the driver to buy some meat from a woman cooking by the station, and this one time I chose not to ask from what animal it came. Pierre relished every morsel with a joy that seemed almost overwhelming, grunting and rolling his eyes heavenward as he ate. When I got out to ease myself, as we say in Cameroon, behind some bushes on the side of the road, a mosquito bit me on my cheek. I only knew it left a mark when Pierre squeezed his fingers through the bars to gently touch the spot on my cheek. This small, subtle exchange, Pierre acknowledging the new lesion on my face, along with my other interactions with him during our long ride to Limbe, narrowed the gap between our species for me in a way that years of academic study and knowledge about genetic similarity could never do. Sadly, this sweet little chimpanzee boy died during an outbreak of pneumonia soon after I left him at LWC. My memories of him and the profound effect our interactions had on me remain vivid so many years later.

I was sad to leave Cameroon. During the several days that Kathy and I spent in the beautiful city of Paris afterward, I pined for the raw, pungent, relative squalor of the African environment we had left behind. Cameroon was life and death; Paris seemed boring by comparison.

And the animal protection and conservation issues in Cameroon, where bushmeat was so prevalent, were compelling for me as an activist. I came to understand that the colloquial term bushmeat refers to any animal killed in the bush and eaten, including chimpanzees, gorillas, monkeys, forest elephants, various species of antelopes, cane rats, pangolins, crocodiles, and turtles. Cameroon law lists chimpanzees and gorillas as completely protected species because they are in danger of extinction. According to the law that was strengthened in 1994, it's illegal to kill, capture, buy, sell, or possess a chimpanzee or gorilla for any purpose, but in 1997, there was almost no enforcement of this law.

Historically, Cameroon's mostly Bantu people had lived in the forest, surviving on what they could kill or gather from their habitat, along with some subsistence agriculture. During the five decades that preceded my arrival in Cameroon, the rapidly increasing human population had urbanized to a greater degree than before. While many people still lived in small rural settlements, those who congregated in towns and cities created the demand for a commercial trade in bushmeat. On its way from the forest to urban dinner plates, bushmeat becomes expensive, as various middlemen, or dealers, who transport and sell it cover their costs and make a profit. I was told that the meat of chimpanzees and gorillas is sweet and succulent, and adding to its desirability in some circles is the belief that eating it increases sexual virility. As chimpanzee and gorilla numbers have dwindled, their meat has become more expensive, adding to its status as a delicacy. Some of Cameroon's affluent people serve it on holidays, at weddings, and to honor special guests.

I loved my contact with the primates, but I was keenly aware that my "opportunity" to know them was a result of the awful tragedy that had happened to them and their families. Chimpanzee and gorilla and monkey families, including mothers with clinging infants, were slaughtered by the thousands for an illegal (at least where chimpanzees and gorillas were concerned) meat trade, and these sweet captive orphans were a side effect. While the sweet old-world monkeys inspired my compassion and the emotionally subtle, mysterious, and relatively gentle gorillas inspired my awe and admiration, the chimpanzees, more than any species I worked with, inspired my deep empathy. I recognized them as my own. I didn't have a definitive plan when I left Cameroon, but I knew I would be coming back. I knew that their cause had somehow become mine.

two

COMMITMENT

Edmund Stone was a British national who had already lived in the United States for seventeen years when I met him in 1997, soon after my first trip to Cameroon. He was producing a talk show for the local Fox Television station when he saw an article in the *Oregonian,* our state newspaper, about my recent trip. He recruited me as a guest on his talk show, and soon afterward we began dating. Edmund was six feet tall and slim. He had a perfectly groomed beard and wore his silky, light brown, shoulder-length hair pulled back in a ponytail. He had first gone to Los Angeles from Britain to work as a BBC correspondent for a radio show about the lifestyles of the rich and famous and had developed a profitable résumé-writing business on the side. Many years later, living in Portland and still writing résumés, he maintained his refined BBC accent, which was quite different from the way people talked in his rural England hometown. He cracked me up with his hilarious imitation of how he once talked. He was good company, and with his tendency toward hedonistic pursuits, such as four-olive martinis after work, he was a welcome addition to my unbalanced, workcentric life.

Later the same year, Edmund accompanied me on a working trip to Cameroon. I had agreed to help British national Chris Mitchell set up a veterinary care program for the new and improved Yaoundé Zoo. Chris was the founder of the Cameroon Wildlife Aid Fund, which would later continue without him, first under the direction of Talila Sivan and the late Avi Sivan, and still later under Rachel Hogan, who oversaw its renaming to Ape Action Africa. I would eventually enjoy the friendship and support of all three of these conservation heroes, but during our trip of October 1997, Edmund and I worked long, grueling hours alongside Chris Mitchell to vaccinate and vasectomize, perform dental work, and tend to various other medical needs of primates at Yaoundé Zoo.

Edmund and I spent two weeks in the town of Limbe, where I had first worked at LWC. Here we befriended three adult chimpanzees, whom I had noticed during my first trip. Jacky, Pepe, and Becky were on display in three small cages located on the back side of the Atlantic Beach Hotel, a quarter mile down the coast from the less expensive Miramar, where we stayed. Their cages of concrete and metal sat about three feet off the ground, lined up in a straight row under some sheltering trees—built there, I supposed, for the amusement of tourists staying at the hotel, should they tire of the beautiful sea on the other side. Edmund and I visited the three chimpanzees and took them food treats at least two or three times a day during our stay in Limbe. Captured as infants by poachers, the chimpanzees had been in captivity most of their lives. Tormented Jacky, a male in his late thirties, was furious and possibly irrevocably insane. Handsome Pepe, a male in his early twenties, wore his loneliness for anyone to see, anyone who paid attention. Sassy Becky, around twenty years old, was the mischief maker, still a kid in some ways. They came to anticipate our visits. They often recognized us when we were no more than silhouettes in the distance, and they welcomed us long before we arrived with excited vocalizations—crescendos of panting and hooting that climaxed in high-pitched screams. "They've seen us," I said to Edmund more than once, and we always hurried our pace so as not to keep them waiting.

As the illegal trade in bushmeat thrived, surviving orphans, too small to offer much meat, could be sold as pets or tourist attractions. Gorilla infants, not as hardy as those of chimpanzees, often lost the will to live and refused to eat after their mothers died. Once they were out of their native forest habitat, they were exposed to bacteria, amoeba, and viruses to which they had little immunity, and many died of infections. It was uncommon to see them in captivity. While many chimpanzee infants died too, either during the hunt or soon after their capture, others survived to languish and suffer for years or even decades in strict confinement on chains or in small cages like Jacky, Pepe, and Becky.

My sense of the injustice of what had been done to Jacky, Pepe, and Becky, and my sadness for the smallness of their lives in these horrible cages, was overwhelming. I pitied them for this fate that was beyond their control, and at the same time I was intrigued by who they were. Edmund and I each developed relationships with the chimpanzees. Pepe stole my heart first and most completely in those first weeks. From the strict confines of his lonely cell, he solicited our interaction with one big arm stretched out through the bars. The first time he beckoned me I closed the gap between us with little hesitation and allowed him to wrap his arm around my back. Somehow, I didn't doubt his gentleness. His muscles were huge and he was fifteen to twenty times stronger than me, but he handled me like a fragile egg. I breathed in his body odor, which was a little like stale sweat but not sharp or unpleasant. I thrilled at being close to him. Although I had read about chimpanzee behavior in natural free-living groups, I hadn't had the opportunity to observe them. I didn't know how to mimic their behavior—to try to communicate with a chimpanzee on chimpanzee terms. On the other hand, living as a "pet" for ten years after his capture, before he was dumped at the hotel, had human-ized Pepe. He wanted to be close to humans, was comfortable around us. I certainly wanted to know and understand him. Our mutual ape-likeness gave us some inherent ground for communi-cation—we understood each other's gestures. At each of our meet-ings, Pepe's first business was to groom me—my face, my head, my arms—and the gentle touch of his big fingers was immensely pleasurable. Free-living chimpanzees spend up to 25 percent of their waking hours grooming or being groomed by their compan-ions. What started millions of years ago, probably when chim-panzees first started living in groups, as a useful activity to control insects and keep clean evolved into an important social activity. Through this intimate touching, chimpanzees establish and main-tain loyal friendships, comfort and calm children, nurture politi-cal alliances, and maintain hierarchies. Maternal grooming from the time of birth imprints the behavior on chimpanzees very early. Even though he was orphaned as a baby, Pepe knew that grooming was an important aspect of any blossoming friendship. By grooming me, he taught me how to groom him—to part his hairs and look for blemishes, dirt, or insects, to scratch or flick his skin as I searched, to move over an area systematically before moving to the next. I used my newfound grooming technique to help cement my friendship with Becky, too, although her recep-tivity to my overtures varied. Sometimes she was happy to spend a half hour visiting with me at the bars of her cage. Other times she kept her distance and turned her back on me, literally. Jacky was a different story altogether. We had learned about his rep-utation from people at LWC before we ever visited the three at the hotel. He had injured several people, and we knew not to approach him. We tossed to him the treats we handed to Pepe and Becky, or sometimes I placed them quickly through the bars of his cage while his back was turned. I noticed that he too vocalized excitedly when we approached. Our visits and the tasty treats we brought punctuated with pleasure the long hours of boredom the chimpanzees endured each day. When the time for our departure approached, I worried about their disappointment—imagined them waiting eagerly for visits that would never come. I didn't want to leave them, especially not knowing when we could return, but we had no choice. We had jobs and companion animals awaiting our return to the United States.

Back home, I thought about the three chimpanzees every day. I continued my advocacy work with IDA and worked some shifts seeing my old dog and cat patients at Pacific Veterinary Hospital, but being an animal activist against institutionalized abuses in the United States was like chipping away at a large stone, and I knew that other competent veterinarians could take over my work with companion animals in the veterinary clinic. In contrast, there was a sense of immediacy and possibility for bringing about big changes in Cameroon, and a sense that, at least where Jacky, Pepe, and Becky were concerned, no one else would save them if we didn't.

I simply could not be happy going about my comfortable life while I was aware that Jacky, Pepe, and Becky remained so bored and miserable. Of course it was true that animals all over the world were suffering as much, or even more, than these three chimpanzees, but they had become my friends. My compulsion to save them was personal.

Edmund felt the same, although perhaps with less intensity. He and I made a promise to each other that we would figure out a way to get Jacky, Pepe, and Becky out of those grim cages at the hotel and give them a better life somewhere else, anywhere else. It was a promise that seems almost casual in hindsight, now that I know what fulfilling it would mean. Although Edmund and I were romantic partners, we were also partners in this mission. Without his mutual commitment to the goals, achieving them would have been truly impossible.

LWC was located on the edge of Limbe very near the hotel, but the center's physical space was limited. It had no facility, or even space for us to build a new enclosure, for three adult chimpanzees. The future of the Yaoundé Zoo was too uncertain for us to think of taking them there. The only way to assure a better life for Jacky, Pepe, and Becky was to start a new sanctuary—a notion that eventually evolved into an actual plan to take the chimpanzees back to some natural habitat forest and protect them within an electric enclosure there. We couldn't move the chimps to a bigger cage somewhere; we would take them back to the forest.

By the time I knew my calling was with chimpanzees in Africa, IDA president Dr. Elliot Katz had come to believe, at least somewhat, in the power of my determination. Enough so that he agreed to continue to pay my salary while I tried to set up a chimpanzee sanctuary in Cameroon. At the same time, he made it clear that IDA was not in a position to fund a project in Africa from its California office. IDA's contribution was to be my salary, and *I* would need to raise funds for the Cameroon project.

Fortunately, I had Edmund on my side. His commitment to me and to the cause, as much as, if not more than, any other factor, gave us our early successes. During the following year, he and I set up IDA-Africa in a room of my house in Beaverton, Oregon, and raised the seed money to start a new sanctuary for the "Atlantic Beach Chimpanzees," as we called them in our fund-raising drive. Neither of us had any experience in fund-raising, but over the course of a year, with the help of friends in our Portland community as well as some from Seattle, we gradually accumulated money through book sales, plant sales, and small receptions. Edmund in particular played the crucial role of reaching out to individual donors. I knew he was driven by his own passion for the cause, and also by his passion for me.

In the spring of 1998, I met French national Estelle Raballand, a pretty, curvy twenty-six-year-old with bright brown eyes and very dark brown wavy hair, which she wore short, just below her ears. She had an easy laugh and spoke near perfect English with a charming French accent. She liked to use American expressions, but she could get them wrong. Her funniest that I can remember: "F you and the horse you wrote it on!" We met at an annual event of the International Primate Protection League in South Carolina. She had lived in the country of Guinea, West Africa, for five years with her American husband, Dana Ward, who worked for a non-governmental organization focusing on family planning and HIV prevention. Estelle had worked with the Guinean government to set up a small sanctuary for baby chimpanzees. She was in the United States temporarily before she and Dana would move with their son, Nicholas, to Yaoundé, Cameroon, for his work. Our mutual friends Peter Jenkins and Liza Gadsby were first to suggest that Estelle and I work together on the ground in Cameroon.

In late 1998, Edmund and I traveled to Cameroon with a plan to stay for a few months to set up the sanctuary. We would then leave it in Estelle's hands and return to the United States to continue fund-raising. Our first goal was to find a sanctuary site, and to that end we used topographical maps to identify areas of interest or we learned about areas during our discussions with other people working in the forests. Sometimes we worked with volunteers from LWC and used their vehicle. Other times we took public transport, renting taxis by the day, to trek through forests of Cameroon's English-speaking Southwest Province looking for a suitable sanctuary site. There were many criteria to consider. We wanted natural chimpanzee habitat where our presence could serve to protect free-living chimpanzees, flat land to make fence building easier, year-round access to the site even during heavy rains, and farming communities of people with the know-how to grow fruits and vegetables to feed the chimpanzees. Location was everything, and finding the right one was essential.

three

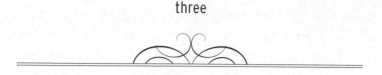

AT THE ATLANTIC BEACH HOTEL

Cameroon's prominent Muna family owned the Atlantic Beach Hotel. The patriarch of the family, Soloman Tandeng Muna, had served as prime minister of West Cameroon, vice president of Cameroon, and then speaker of Cameroon's General Assembly for fifteen years. His six sons and one daughter were all successful in one profession or another, and the family name was well known throughout Cameroon. In 1998, George Muna, a businessman based in the city of Douala, took over management of the Atlantic Beach Hotel and two other hotels the family owned in Limbe. His move to the coastal town was precipitated by the departure of the family's French hotel manager, who had fled the country with money extorted from the hotel business.

Before his departure, the French manager had assured us we could take Jacky, Pepe, and Becky to our sanctuary, but we weren't sure what his exodus would mean for the chimpanzees. Now, we needed to get the approval of the Muna family firsthand. Because Estelle and Dana had moved to Cameroon a few months before Edmund and I came back in November 1998, Estelle first went alone to meet with George Muna. To our great relief, George was quick to agree that we could and should take the three chimpanzees away from his hotel.

I met George that November, soon after Edmund and I arrived from the United States, when I approached him in the wood-paneled hotel reception area about the possibility of renting an oceanside house owned by the Muna family. The neglected two-bedroom house, white with blue trim like the Atlantic Beach Hotel located one hundred yards from it, was perched idyllically on a low rocky cliff above the ocean. Edmund and I needed a base of operations, and I wanted to live in this little house badly enough to come knocking on George's door, so to speak. George was slightly overweight and not exactly handsome in a classic sense, but his charisma filled the room and made me self-conscious during our first meeting. He was wearing a dark blue dress shirt and maroon tie, and he spoke to me with the dominant self-confidence of Cameroon's aristocratic class. After I introduced myself and explained my purpose, George led me into his office, past at least ten people who waited outside his door for an audience. I would learn that Cameroonians wait far more patiently and graciously than Americans and that receiving important people first is normal. Businessmen and government officials generally scheduled their day's appointments with everyone arriving at the same time and then received each person in an order appropriate to his status. It took me years to understand that this inefficient system that wasted so much of my and everyone else's time was the only one that could work in a society with so many communication, transportation, and general infrastructure challenges. No one could be expected to "keep the time" for precisely scheduled appointments. George sat in a leather chair behind his big wooden desk, and I sat on the edge of a hard-backed chair in front of it while he explained why he was grateful to us for proposing a solution to his chimpanzee problem.

George Muna and I spoke easily to each other, which was surprising considering our vastly different backgrounds and perspectives on the world. My white liberal guilt, vegetarianism, and casual clothes were in stark contrast to George's status consciousness, culturally entrenched carnivorism, and tailored suits.

"The chimpanzees were here when my family took ownership of the hotel, and they've caused us trouble from the beginning," he told me.

Rising briefly to pull a cardboard filing box from a room-length shelf that contained dozens, he spoke of insulting letters he had received from European hotel guests complaining about the living conditions of the chimpanzees. Sitting again, he pulled a letter from the box and handed it to me. It was in French, and I couldn't read a word, but scanning the first page and glancing at the second, I nodded knowingly, trying to look sympathetic about the problem of his family's business image with French-speaking Europeans.

George went on to say that his bigger concern was about safety. "I worry every night that those chimpanzees will hurt a hotel guest," he said. "They've already wounded several employees."

"That's a valid worry," I affirmed sincerely.

While our motivations for wanting to relocate the chimpanzees were different, George and I could certainly eye the same prize of getting them moved. It was the foundation for an alliance of sorts. Before I left his office, George had agreed to rent Edmund and me the little run-down oceanside house, in the most beautiful location I had ever lived, for the equivalent of $100 per month for three months.

Edmund and I could hear Jacky, Pepe, and Becky vocalizing from the house. When we weren't out in the field looking for a sanctuary site, we visited the chimpanzees frequently. Going to see them was the first thing I did each morning, even before coffee. I hardly ever ate without saving something for them, although they got enough food from the hotel staff, and we were always bringing gifts of fruit juice, tennis balls, which Becky and Pepe liked to toss back and forth with us, clip-on key chains, magazines, mirrors, and anything else we thought might ease their boredom. Becky especially enjoyed the magazines we gave her, and she paged slowly through them, looking at the pictures. Pepe loved the plastic mirrors we brought—most animals cannot recognize their own reflections, but chimpanzees can. Pepe opened his mouth wide to examine his teeth and held it behind himself to examine a scratch on his lower back. He even used it to surreptitiously watch what the other chimps were doing behind him.

Jacky, the oldest of the three, was in the middle cage. We knew he had been at the hotel since the late 1960s, possibly earlier. People called him the "mad chimpanzee" of Limbe, and it wasn't difficult to see how he earned that reputation. He refused to make eye contact with us, and his various forms of stereotypy, while heart wrenching, did make him appear lost to the sane world. In one of his most disturbing and frequent manifestations, he placed one open hand in his mouth while rapidly and forcefully pounding the top of his head with his other fisted hand. He abused himself like this frequently and for minutes at a time. Jacky was bald on top of his head, and the skin there was thickened, as a result of his head beating. Other times he sat with a fixed upward stare as he rocked back and forth for hours, sometimes masturbating while he rocked. He was unresponsive to the taunts of human onlookers and seemed oblivious to their presence until the fateful moment when a careless person ventured too close to his cage and paid a high price for the mistake. With lightning speed and certain intent Jacky could grab hapless hands, pull them into his cage, and with a single bite inflict irreversible damage. Restaurant employees, careful not to get too close, threw his food to his outstretched hands.

One evening after we ate dinner in the hotel restaurant, Edmund and I were visiting the chimpanzees just after dark. I was filling bottles of drinking water for the chimps from a freestanding tap when I heard Edmund's terrified scream for help from behind me. I spun around to see him plastered against Jacky's cage, pulling back futilely against the powerful iron grip on his right arm. When Jacky decided to let go, Edmund's own strength propelled him backward and hard to the ground. Before I could get to him, he had rolled over to a kneeling crouch, cradling his hand in pain. During the interminable several seconds that I couldn't see his hand, I feared the worst. I pulled on his arm, trying to see his right hand, which he stubbornly enclosed in his left, frozen by the pain.

"Let me see your hand!" I shouted. When he finally opened his left hand to reveal his bloody right one, we saw together in the dim light that all his fingers were intact. "Thank God," I said. While he still clutched his hand in agony, I was grateful he hadn't been permanently maimed.

One mile away, at the veterinary room of Limbe Wildlife Center (we had to take a taxi to get there because we still didn't have our own vehicle), I stitched a deep bite wound, an inch and a half long, and left open another small but deep puncture. Edmund's right hand and all its fingers would eventually return to full function.

Other people who had been hurt by Jacky were employees of the hotel. Three painted words, MONKEYS ARE SEIZING, on each end of the cage warned visitors to stay away, and it seemed that tourists took the warning seriously. Cameroonians were typically afraid of large animals anyway, but employees were injured because they were doing work around the cage, or because with familiarity, over time, they became too casual in wandering near it. At least two of them lost fingers, and one was left with a permanently frozen knuckle. The terrible injuries caused people a lot of pain and were financially costly for the Muna family, but Jacky could have caused much more damage than he did. With a few well-placed chomps he could have removed a human hand, but in the cases of which I was aware, he bit only once before letting go.

We wondered whether Jacky could ever be socialized with other chimpanzees. If he were to escape our enclosure at the sanctuary and severely injure or kill someone, not only would it be a horrible and arguably preventable tragedy, but it would endanger the whole project. One consultant told me that euthanasia was the only reasonable solution for Jacky. For me, ending Jacky's life would have been unethical, and leaving him at the hotel alone while we took Pepe and Becky to a better life would have been unfathomably cruel. Jacky was a victim of a terrible injustice, and I wanted to make it right for him. Although I worried about the dangers of taking Jacky to the sanctuary we would be building and went so far as to get some Prozac donated for him (which I never gave him), I never considered any option but giving him a chance to recover at the sanctuary. Edmund and Estelle were my allies every step of the way in this decision. We didn't know of any precedents where chimpanzees as psychologically damaged as Jacky had become well-adjusted members of a chimpanzee society, but we knew we had to try.

Pepe and Becky were much easier to love initially. The two of them had been together since they were babies, raised as pets by a French expatriate couple. Although the couple probably didn't start off with terrible intentions, they surely had very limited understanding of Pepe's and Becky's needs. No human home and garden could accommodate normal chimpanzee behavior. When the chimpanzees became adolescents, much stronger than anyone in the household and difficult for the couple to manage, they were left at the Atlantic Beach Hotel, where Jacky already languished in the middle of three cages. Other chimpanzees had lived and died there many years earlier. Pepe and Becky were placed in individual cages on either side of Jacky—deprived of the ability to touch or comfort one another with mutual embraces. Even if anyone thought of it, which they might have, the small mercy of putting Pepe and Becky in cages next to each other couldn't have been accomplished without the ability to anesthetize and move Jacky. The three chimpanzees had lived in their single-file configuration for ten years when we met them in 1997.

Pepe always loved my attention. Back and shoulder massages became part of the bargain of my friendship with him. After he groomed me, it was his turn. He turned and plastered his back to the wall of the cage and dropped his arms to his side. I massaged the strong muscles of his arms, neck, and upper back just as I would have liked it, and when my fingers got tired, I groomed his skin using the technique I learned by watching him. Pepe loved my touch and seemed to really enjoy his massages, which I delivered lovingly, but writing this now, so many years later with so much more experience, I can't help but wonder if he would have preferred simple grooming. Whenever I pointed to Pepe's hand, he knew instantly to pass his huge extremity through the bars, and my small hands cradling one of his were dwarfed by the comparison. I cleaned under his fingernails and toenails, cutting them with my large toenail clippers when they got too long. Pepe was always tolerant and trusting.

Estelle warned me to watch him carefully for sudden changes in his temperament. Adult male chimpanzees are emotionally volatile, pushed around to a large extent by their hormones. In the wrong set of circumstances, all can be deadly dangerous. Pepe was locked in a cage behind a hotel, while I was free to come and go. While he himself should have been living free with his own kind in a forest far away from us, he was reduced, in his lonely boredom, to begging for interaction from me on my terms. These were definitely the wrong circumstances. The sadness in the situation was ever present for me. The danger was brought home when Pepe bit a long, deep gash into the upper arm of a volunteer from Limbe

Wildlife Center. I wasn't there at the time and never knew the exact circumstances under which the bite occurred, but the man had worked with chimpanzees for years and knew how to behave around them. It was a disturbing incident. Nonetheless, during my sporadic long visits with Pepe in those early years of 1997, 1998, and 1999, the gentle side of this large and powerful chimpanzee was the only side he showed to me.

Becky, on the other hand, was clever and often sweet natured, but also mischievous and prone to mood swings. She had shiny, almost blue-black hair and a beautiful black face, from which her sharply contrasting topaz-brown eyes sparkled.

One morning when Edmund was hard at work on repairs to our apartment, I was walking home after a meeting with a local businessman from whom I was soliciting donations of metal for the sanctuary. When the chimpanzees saw me passing, they extended their familiar excited greeting of pant hoots and screams. I was wearing a light beige dress that buttoned all the way down the front and fell to within a few inches of my ankles—the only dress I owned that was fit to wear to business meetings with government officials or potential contributors. Ordinarily I would have changed clothes before visiting the chimpanzees, but I couldn't resist their hearty invitation this morning. Something in Becky's bright-eyed expression of anticipation caused me to visit her first.

As I reached the side of the cage where she sat watching my approach, I realized she was interested in the half-full bottle of drinking water in my right hand. Her eyes traveled from the bottle of water to my eyes and back to the water—a master of efficient communication. I took the screw cap off, although she would have had no problem doing it herself, and handed her the bottle. She took a big swallow and set the bottle upright on the concrete floor. Careful not to tip over the water bottle, she scooted close to the cage bars and rotated her leg to reveal a spot on her wide inner thigh that she wanted me to see. As she pointed to the area, I reached my forearms through the bars to examine it.

As Pepe had done with me, and as I had watched other chimpanzees at LWC do with one another, I rhythmically opened and closed my mouth with some lip and tongue smacking to assure Becky I only intended to groom her. I did my best to mimic what I had seen, but I've learned since then that the almost universal chimpanzee mouth movement during grooming is highly individualized. It can be subtle and quiet or vigorous and loud depending on who is doing it. Regardless of the exact characteristics, it communicates benign intent with perfect clarity. Even chimpanzees who have been orphaned as young infants tense their lips in concentration and pair grooming with mouth movement as they mature, which implies an innate neurological connection between the nerves used in precision hand movements and those innervating the mouth. My own mouth positioning when I perform a difficult surgery or other truly challenging fine motor task with my hands is of the same quality, although more subtle, as that of the chimpanzees when they concentrate on grooming someone. For us it may be a benign and useless remnant now, but the connection may have played a role in the evolution of human language.* Grooming the chimpanzees didn't trigger my unconscious mouth positioning the way performing surgery on them would, the way grooming me did for them. My mouth movements as I groomed Becky were practiced and intentional, but she understood my meaning.

With her head bowed she watched my fingers part her hairs to reveal a tiny inflamed pink spot—it looked like an insect bite. I flicked at it gently as I raised my eyes to hers reassuringly for just a

*G. S. Waters and R. S. Fouts, "Sympathetic Mouth Movements Accompanying Fine Motor Movements in Chimpanzees (Pan Troglodytes) with Implications Toward the Evolution of Language, *Neurological Research* 24(2) (March 2002): 174–80.

moment and then proceeded to groom all around the tiny lesion on her thigh. Soon, she too began grooming her own thigh, so the two of us occupied ourselves with this single important task.

In the cage next to Becky, Jacky pretended to ignore me, but I was aware of Pepe watching and waiting a few yards away in the cage on the other side of Jacky's. I needed to visit him, too, and I wanted to find water bottles to give both Pepe and Jacky in case they were thirsty. Even if they weren't, I always tried to be equitable in my distribution of gifts, however small, to the three of them.

Just as I was preparing to leave, Becky casually grabbed a fistful of the ample beige cloth of my dress and pulled it through the bars.

"No, Becky!" I said with alarm and to no effect. Without a hint of real malice, but with unalterable intent, she used both her hands to grab my dress from the bottom. Within moments, three-quarters of my lovely dress was crumpled on the dirty concrete cage floor in front of Becky, and I was struggling to keep the back of the dress stretched down over my behind. At first I thought Becky's primary motive was to keep *me* with her, but her loving caresses of the dress soon convinced me that she really just wanted to share the special cloth. I had two choices: get out of my dress and leave it with her, or convince her to release the dress and me. Slipping the dress over my head and running in my tattered underwear past the hotel lobby and restaurant, where people had begun to arrive for lunch, was a very unappealing option. Glued against the cage, I had nothing with which to bargain, no way to buy back the dress. I tugged gently on the cloth.

"Becky, please give it back," I begged. Clutching the dress firmly in her left hand, she picked up the water bottle with her right and poured water from it onto the dirty concrete in front of her. After a brief glance at my disapproving face, she set the water bottle down so she could use both her hands to scrub the floor with the never-to-be-beige-again cloth. As she cleaned, she stuck her tongue in her cheek as she sometimes did when she was busy, and miffed as I was over the ruining of my dress, I thought she was adorable.

As I considered the hopelessness of my predicament, I saw George Muna walking some distance away along the driveway of the hotel with two business associates. I shouted his name, trying to sound calm so as not to upset Becky. She had never been aggressive with me, had never threatened to hurt me, but as I was in a vulnerable situation, I thought it prudent not to irritate her. George waved casually and kept walking. Not a big help!

Fortunately, after several more minutes, George came back. When he got just close enough to hear me easily, I asked him to bring me a French red apple from the restaurant so I could try to trade it to Becky for my dress. Expensive red apples imported from France were a delicacy for Becky, and I had seen some in the restaurant that morning. While I waited for the apple, I planned my strategy. I didn't think Becky would try to force the apple from my hand, but just in case she didn't respect the rules of fair trade, I planned to ask George to put the apple on the ground near me, out of Becky's reach. When Becky released the dress, I would deliver the apple to her. I thought she would easily understand the bargain I intended. But when George returned carrying the apple, he stopped on the far side of the cage, out of Becky's reach, and held up the apple temptingly for her. When she saw it, her face lit up. After thinking about it for no more than a few seconds, she dropped the dress, crossed the cage, and reached her arm through the bars for George to throw it to her. Instead he laughed and refused, thinking it funny that he had tricked her into releasing me.

Obviously, his sense of fairness did not extend to chimpanzees. Both Becky and I were furious at his outrageous act of betrayal. She barked and spat at him, while I yanked the apple from him and handed it to her. Eventually, as George's many other acts of kindness, some of which helped chimpanzees, overshadowed this injustice against Becky, I forgave him. Becky never did. Thereafter, whenever she saw him, she glared with malice, and I knew he better never wander within her reach.

These chimpanzee visits were happening when we weren't exploring a forest, or meeting with villagers, or meeting with people in other organizations, trying everything we knew to find a suitable sanctuary site. We spent more time discovering sites to explore than we did actually exploring the sites. Every site we looked at had a fatal flaw—it wasn't accessible in the rainy season, or it had no road access at all, or the politics were too complicated.

After three challenging months, during which we did not find our perfect sanctuary site and became seriously worried about whether the money we had raised would be enough, Edmund needed to go back to work in the United States. We decided that he would lease out his house in Oregon and move into mine, to save money and focus on raising funds, while I stayed in Cameroon to find a site and set up the sanctuary. I had never planned to stay in Cameroon after Edmund left, but for me to leave would have meant failure, would have meant leaving Jacky, Pepe, and Becky in their cages at the hotel, probably for the rest of their lives. I wasn't willing to fail.

Communication, both within Cameroon and internationally, was very difficult in those days before cell phones and reliable e-mail. Landlines were few and far between, and connections were erratic. I knew my communication with Edmund would be limited, and I was sad to see him go, but his crucial fund-raising in the United States would enable Estelle and me to push forward in Cameroon.

I loved Edmund, but after he left, with the distance and loneliness, we began to grow apart. The relationship would change very gradually. In the end, it would be me and my needs that would bring about the end of the romantic relationship, not him or anything he did.

With all my heart I was dedicated to the mission we had under-taken. I knew that Jacky, Pepe, and Becky were not the only cap-tive chimpanzees living in small cages in Cameroon, and I knew that we would rescue others. My idea was to build a small infra-structure with one or more forested enclosures for older captive chimpanzees. I would work with Estelle, other volunteers, and a local staff to set up the sanctuary and then direct it from afar, coming and going from Cameroon as necessary. The problem with my vision at the time stemmed from a very limited under-standing about the scope of the chimpanzee orphan problem in Central Africa and about the difficulty of saving and caring for chimpanzees in a country like Cameroon. Ultimately my love for Jacky, Pepe, Becky, and others would inspire a stronger commit-ment, by far, than any I had ever made. I could not yet fathom the extent to which it would change the course of my life.

four

MEAN STREETS

The big city of Douala, considered the business capital of Cameroon, notorious for its high crime rate, lay between Limbe and Yaoundé where Estelle lived with Dana and their son. One morning after leaving Limbe en route to Yaoundé, I took a detour off the main road through Douala to check e-mail at a cybercafé in the Akwa District, near Douala's port. As I approached the door of the business—rather preoccupied as usual and not expecting any surprises, since I had been there several times—I could have tripped over a grisly surprise that literally took my breath away. About three yards from the cybercafé threshold, and equidistant from where I stood with my mouth gaping open, lay a bloody, charred human corpse. Blood spattered the concrete under my feet, and the odor of burning flesh was thick in the air. My eyes instinctively avoided what had been the man's face and fell instead upon two military police officers who chatted casually several yards away. Turning my back on the corpse, I peered through the glass of the cybercafé to see people busy at computers. Business appeared to be going on as usual, and although it was a shocking affair to come across, the murdered man on the sidewalk hadn't negated my need to check e-mail. I entered the cybercafé and glanced around for an empty computer. Long, narrow tables made of plywood that was painted white were positioned along two walls of the room, which was also painted white and measured about fifteen by fifteen feet. Two similar tables placed end to end divided the room down the center. Pieces of vertical plywood about two by two feet divided the tables into narrow computer stations. Seeing that two computers were vacant, I paid the female attendant, who sat at her own small desk by the door, for an hour of Internet time. She handed me a small piece of paper with a password written in blue ink and spoke to me in French as she gestured to the computer I should use. I settled into my station, took a couple of deep breaths to calm my racing heart, and glanced at the man sitting at the station next to me. His hair was very short, and he had a light complexion for an African. It was hard to tell his age, maybe early thirties. He wore jeans, a Polo shirt knockoff, polished brown loafers, and wire-rimmed glasses, which were in good condition. I took his appearance to be a sign of relative affluence—that and the fact that he was sitting beside me in one of Douala's more expensive Internet cafés. To use the computer here cost the equivalent of two dollars per hour.

In 1999, Internet connections were still slow and spotty. The man had pushed his chair back from the table slightly. He seemed to be having a moment of downtime, and my quick glance over the divider at the little hourglass on his computer screen confirmed that he was waiting for his Google search to process.

"Bonjour," I said with a slight nod. "Good morning," he shot back effortlessly, aware from my accent that French wasn't my language and letting me know as he turned slightly to face me that he was willing to talk. We exchanged names and pleasantries. His name was William, he taught at the Anglophone university in the town of Buea, and he was visiting his brother in Douala. He knew details about the deadly drama that had unfolded outside and was willing to share them. Like many Cameroonians, William was an eloquent storyteller.

"It's a story of criminals who got caught by the people," he summarized, before continuing with the details. "A woman came out of the Standard Chartered Bank, a few blocks away, and flagged a taxi. A minute after she entered the taxi, the driver picked up two more passengers."

Taxis in Cameroon are usually shared—drivers pick up as many people as they can who are going the same direction. To flag a taxi in Cameroon's cities, a hopeful passenger simply stands by the road and looks in the direction of any approaching taxi that isn't crammed too full. She might lift her arm forty-five degrees from her side with her index finger extended. As the taxi slows, the flagger shouts her destination through the open window. If the driver intends to accept the new passenger, he simply idles in place to allow time for her to crawl in, or if he is blocking traffic he might pull off to the side of the road toward her, or he might even beep his horn—always a cheerful sound of acquiescence to the flagger on the street. The taxi driver's refusal is made clear if he simply keeps driving. It wouldn't have been unusual for the taxi driver who picked up the woman from the bank to pick up other people as well. In itself that wouldn't have been cause for alarm.

William continued. "Soon the victim realized she was in the company of three collaborating bandits, including the taxi driver. She managed to get out of the taxi, but she didn't manage to take her purse with her."

I interrupted him. "Did they force her from the car and keep her purse, or did she escape from the car?"

"This detail isn't clear to me. What we know is that she landed on the hard concrete sidewalk, and she still sat there as she pointed after the taxi yelling, '*Thieves! Thieves!*'"

Because Cameroon's unemployment rate is very high, most people have been victims of crime at one time or another. The legal system rarely offers any justice for victims of property crime, so hatred of thieves in the general population is particularly intense, and vigilantism is common. Shouting "thief!" in a crowd is a good way to get someone killed.

After giving me a few moments to absorb what he had told me, William continued. "After the victim fell out of the taxi and sounded the alarm, a growing crowd of angry citizens began running alongside the taxi. When the car had to stop in traffic, the criminals had no escape. The people pulled them out and proceeded to beat them to death," William finished matter-of-factly.

"Where are the other two bodies?" I asked, trying not to sound as shocked as I felt.

William stood, and I followed suit, as he pointed to two dark spots where the bodies had previously lain on the concrete walkway. "Their families have already collected them for burial. The military police are waiting for the last one to be claimed," William told me.

"Were they already dead when they were set on fire?" I asked hopefully, sinking back into my chair.

"We can't know," William said with a shrug, as he too sat down and turned his attention back to his screen.

"It's really shocking and sad for people to be killed so brutally on the streets"—I tried to maintain an unemotional tone as I stated what I thought would be obvious—"without benefit of a trial or anything."

When William didn't respond, I queried, "Don't you think so?"

"Well, criminals should think twice," William said. "At least the victim got her purse back." He was busy typing at the computer again.

"That much is good," I said lamely. Our perspectives were clearly different.

The incident held the distinction of being the first time I saw vigilantism at work in Cameroon. I paid for a second hour of computer use, taking my time, staying long after William left, hoping that the corpse would be removed before I would have to leave the cybercafé. I tried to answer e-mails, including one from Edmund, but the violent vignette played over and over in my head. I tried to process the different elements of the tragedy—the tinderbox fury of the crowd, the unimaginable suffering of the men who died, the pain and anguish of the family members, especially the parents, who claimed the bodies, the callous resignation of bystanders like William. It seemed that poverty and hopelessness, born of years of institutionalized corruption at every level of society, were a toxic mix in this country that held so much beauty. Finally, I had no choice but to rush past the body again. I would join Estelle in Yaoundé to continue our search for a sanctuary site, a place in the forest where I expected life would be far less perilous, at least for humans.

five

THE MBARGUE FOREST

O ur search for a sanctuary site had taken Estelle and me out of the Southwest Province and through some beautiful forests in South and Central Cameroon, but our travels served to illuminate how well our mission could have been served by a larger budget—money to build a bridge or a short road could have allowed us to create access to some appropriate but inaccessible sites.

Then Karl Ammann, whose photographs were educating the masses outside of Africa about the horrors of the bushmeat trade, had introduced us to Jean Liboz, the French national who was a director of the Coron Logging Company in Cameroon. Both Liboz and Karl thought the Mbargue Forest, near the eastern boundary of the Central Province, which was in Coron's logging concession, could be a good place for a chimpanzee sanctuary. They especially liked the idea that our presence in the forest could provide some protection for the small populations of free-living chimpanzees and gorillas that remained there.

Now, en route to the Mbargue Forest for our second visit, Estelle and I were responding to Liboz's invitation to sleep for a night in the luxurious hilltop logging camp of the Coron Company. From the camp, built in a clearing on the tallest hill, the multiple shades and textures of the vibrant forest fell away in all directions and as far as the eye could see. The uniformly red-brown village of Pela at the foot of the hill and the red-brown snake of road leading to the camp appeared like wounds inflicted to the luxuriant green. The ten comfortable sleeping quarters of the logging camp were in five prefabricated metal modules of white—each module had two bedrooms with a toilet and shower between them. The modules were lined up along a covered walkway of polished hardwood slats, which connected them to the spacious open-air den and dining room. Modern plumbing throughout the camp was gravity fed from a two-thousand-liter storage tank on tall stilts, which in turn was supplied by a deep borehole well. A muffled generator purred continuous electricity. French cuisine was prepared and served with fine wine by the Cameroonian cook in the dining room. The environment was a stark contrast to the surrounding villages, with their mud huts, cooking pots in open fires, and lack of running water and electricity.

If we were to choose the Mbargue Forest for the sanctuary, Liboz was agreeing to clear our driveway and campsite with the bulldozer of Coron. Equally important, their logging trucks would transport our metal and other supplies over the 225 miles of mostly dirt road from Yaoundé to our sanctuary wherever we chose to locate it within the Mbargue Forest. I had very little money to work with, and I knew Liboz's contribution could mean the difference between success and failure.

The Cameroon government had also pointed us toward the nationally owned Mbargue Forest. The provincial delegate of the Ministry of the Environment and Forestry (MINEF), the branch of government with which we would necessarily collaborate, singled out the area as one deserving of development. President Paul Biya's wife was from Nanga Eboko, a town in the same district, and the local people were strong supporters of the president. It was about politics for them. For me, choosing the Mbargue Forest could be a bow to logistical necessity and political expediency. I really hoped it would meet our criteria.

I never heard Jean Liboz called by his first name, and I never called him by it. He was about my age, but he was Monsieur Liboz or simply Liboz to me, depending on whether or not he was present. To refer to him differently now would be disingenuous. It does not reflect a low regard for the man, but rather a lack of casual familiarity. The language barrier and the short amount of time we were ever in close proximity to each other ensured that our personal conversations were strained and brief. Liboz was a swaggering and handsome roughneck, whose ever-present holstered pistol was visible or hidden, depending on the situation. After decades as a logger, he seemed to be having a twinge of conscience about the role of logging in the demise of apes and other species of wildlife, but the truth is that I never fully understood Liboz's motives for donating his time and company resources to helping our very local and obscure efforts for chimpanzees. He never took photos of our smiling and grateful faces, never publicized the assistance he gave us to promote a greener image for his company. He seemed to genuinely care about wildlife. Perhaps he *was* motivated by guilt. If so, he didn't seem to harbor any similarly soft sentiments toward domestic animals.

We had first visited the Coron logging camp two weeks before, just after Liboz had driven us from Yaoundé for a one-day whirlwind tour around the Mbargue Forest, along the logging road he was still completing. The 225-mile road we took from Yaoundé to the Mbargue Forest wasn't paved, and he had driven like hell over the rough dirt roads, flying through roadside villages, hitting a chicken, a pregnant goat, and finally a baby pig along the way. The sight of a flopping, then suddenly limp, baby pig in the rear window, following shortly after the other unnecessary vehicular animal carnage, rendered me nearly hysterical in the backseat.

"He's driving like a lunatic, *Estelle*!" I spurted at her, like it was her fault. They were both French, after all. "I can't take it anymore. Either he quits killing animals, or I want out of the damn car!" I drew my line emphatically, although I knew getting out of the car on a rural road in Cameroon wasn't a viable option. I also knew that Liboz couldn't understand my English, so my speech wasn't entirely reckless.

"Les animaux . . . une docteur veterinaire . . . une vegetarienne . . ." I understood some of Estelle's rapid-fire French descriptions of me as she tried to explain my sensitivity to a confused Liboz. I knew that she too was upset about him hitting the animals, and quite happy to have me to blame for the intervention. Liboz slowed down a little in acknowledgment of my feelings. Although he only maintained the slower speed for about a minute, he used his brake more liberally afterward, meeting my eyes once in his rearview mirror when he did so to spare a goat.

As my blood pressure normalized, I mulled over the fact that Liboz's thoughtlessness wasn't only toward the animals, but also toward the people in the villages who had so little and who relied on these animals for their livelihoods. I imagined how they felt watching three white people fly so recklessly through their villages. Few white people passed through here, and Liboz was doing his part to establish our bad reputation, I feared.

This time Estelle and I had come in the little blue 1990 Pajero that I had bought for the project a month earlier when I couldn't find a pickup we could afford. I pampered it across the bumpy washboard road and took twice the time to drive from Yaoundé that Liboz did. I was delighted to have wheels after taking public transportation for five months, and I was still too naive to realize just how inappropriate the small SUV would be for the construction work we would be doing in months and years to come.

That night over cocktails, through Estelle's translations, I heard Liboz speak of his efforts to minimize the impact of his company's activities on wildlife. He prohibited his employees from transporting bushmeat in or on his company vehicles, although he knew his policy was ignored in his absence. Even having the policy was somewhat progressive for the time. He told us that he had discovered his employees transporting the body of a forest elephant across the Sanaga River on the company ferry, which he himself had built, and he dumped the dead elephant in the river, depriving everyone involved of financial benefit. Involving Cameroon law enforcement in a wildlife issue—with the inherent delays in production and the police expenses he would be expected to cover, along with local political problems that would stem from it—wasn't something Liboz would have considered.

He admitted little culpability for habitat destruction. "Coron takes an average of only one tree per hectare (about 2.5 acres)," he told us. I knew the number sounded more benign than the practice it described. Each tree the company cut likely brought down others on the way down, and the loggers caused a lot of damage to the forest as they accessed and removed cut trees. Nonetheless, Liboz argued that encroachment by village farmers was much more damaging to habitat than logging.

"Overpopulation is the real problem. Local farmers are the problem," he insisted.

I could see his point. The last few months had provided me with a fast-track education about Cameroon's conservation issues. Estimated birth and death rates indicated that the population here was probably increasing by about four hundred thousand people per year, while job opportunities were not increasing. I had seen firsthand that people were pushing farther and farther into the forests, hunting and farming to survive. Staking a claim to Cameroon's "free land" offered a better quality of life for many than trying to survive in urban areas without jobs or government assistance. From a conservation perspective, the burgeoning human population was a big problem, but new human settlements often followed the logging roads into the forests. The roads sliced into pristine forests for which inaccessibility had once afforded protection, and some of the people who eventually settled along the roads came first to work in the logging companies. Others came to hunt and provide meat for the logging company employees. The logging roads came first, then the new settlements and the slashing and burning for subsistence agriculture that followed. Without the logging roads, I speculated that Cameroon's people wouldn't have spread out so much, would have chosen to live closer together in larger rural communities where they might have relied more on farming than on hunting. To be sure, my compassion extended to the Cameroonian people I was meeting, but my perspective was that of a conservationist who saw huge advantages in keeping wildlife habitat inaccessible.

The increased forest access provided by logging roads also dramatically increased the pressure from commercial hunters, who used the logging roads, and often the logging trucks, as conduits to transport bushmeat to urban centers. These poachers were driving species extinct, rendering forests silent, and creating primate orphans in need of sanctuary. I kept quiet, though. Self-serving logic aside, Jean Liboz was a gracious host to us and was trying to do something good for conservation. I didn't see any advantage in annoying him, and I guess Estelle agreed because she was uncharacteristically diplomatic in the conversation, as far as I could tell.

During much of the three or four hours we visited with Liboz that night in the logging camp, he and Estelle chatted away in French. After a while, she quit bothering to translate unless their conversation concerned our work. It was boring for me, and the constant noise of their unintelligible chatter was nerve-racking. My lack of French seemed a ridiculous obstacle, and it was extremely frustrating for me. We had moved out of Cameroon's Anglophone area to begin looking for our sanctuary site in the Francophone zone a couple of months earlier.

In the late 1800s this geographic area of Central Africa was colonized by the Germans, but British and French troops forced them to leave during the First World War. In 1919, soon after the war ended, the former German Kamerun was partitioned, with France gaining administrative control over the largest share and the United Kingdom getting a smaller section. French Cameroun achieved political independence in 1960, and in 1961 a UN-sponsored referendum allowed the British Cameroons to decide their fate. The southern third voted to unite with French Cameroun to form the Federal Republic of Cameroon, while the northern two-thirds voted to join with Nigeria. A new constitution in 1972 changed the name of the young country, with its eight Francophone and two smaller Anglophone provinces, to the United Republic of Cameroon.* While the government claims both French and English as its official languages, and many *urban* schools now teach both languages, it is still very rare to find anyone who speaks English in the rural Francophone provinces of Cameroon.

I had been trying to pick up the French language, but at the age of forty, the language acquisition part of my brain seemed to be completely atrophied. I was trying to learn from Estelle, sometimes working on pronunciation of words during our long rides, but French wasn't coming easily to me. I would have excused myself from the conversation with Liboz, but I had run out of books to read. At least I was enjoying the gin and tonic, and the cool night air in the open hilltop den was breezy and pleasant. I managed to stay lost in my thoughts for much of the time, but at one point I heard Liboz say "Americaine," and I knew he was referring to an American woman or women.

"What did he say?" I asked Estelle.

* BBC News, Africa, Cameroon Profile; www.bbc.co.uk/news/word-africa-13146029.

"He said French women are much prettier than American women," she told me without hesitation and then cracked up laughing at my openmouthed, stunned expression. Evidently, her newfound diplomacy did not extend to me. I chose not to interrupt them again, and we all went to bed soon afterward, as Estelle and I planned to get an early start the next morning.

There were sixteen small villages around the Mbargue Forest. After stopping briefly in two of them, we arrived by noon in the tiny village of Bikol. Although surrounded by the lush green forest, Bikol itself was another brown village. The houses, the earth upon which they stood, and the new logging road passing through the village's center were various shades of reddish brown. No grass, which could conceal snakes and scorpions, was allowed to grow in the village.

The infrastructure of Bikol consisted of twelve small rectangular houses, the walls of which were made of sticks and mud. Rain had melted a twenty-inch section of mud plaster from the front wall of one of the houses so I could see its frame—a latticework of wooden and bamboo poles laced together with strips of tough plant fiber. The roofs were made of raffia palm fronds woven together into long shingles. The land and its forest had produced every building material. Like thousands of village houses we had seen in rural Cameroon, they each had one or two rooms with swept dirt floors. Inside some of them, I knew that women would cook over open fires without chimneys, creating a haze of choking wood smoke that would exacerbate respiratory problems in their children.

We drove slowly through the village, careful not to hit the chickens and goats that lingered casually in the new road and perceived no threat from our approaching vehicle. The village looked mostly deserted. Just when it seemed that we would pass through without seeing anyone, we spotted four men sitting on low bamboo benches around the smoldering embers of what had been a small fire. A roof of woven raffia palm fronds shielded them from the sun. We stopped the Pajero and got out. Two of the men stood and approached us. A handsome man of medium build who looked to be in his fifties introduced himself as Chief Gaspard. He introduced the other man, who was slighter, younger, and missing several of his front teeth, as his brother Colbert. They both wore loose, ill-fitting Western clothes.

They spoke to us in French, a much slower cadence than Estelle's. Everyone in Cameroon has a local language that they learn at home as their first language. The country has many different ethnic groups and more than 270 local languages. People who live only a few miles from one another can speak different local languages. Although the country claims French and English as its two official languages, French is the language taught in rural schools. Most people who have attended even a year or two of primary school speak basic French. Very young children, orphans who didn't go to school because they had no one to pay for it, and many older men and women speak only their local language.

The two men invited us to join them under their raffia roof, gesturing for us to sit on one of the three benches that formed an incomplete square around the slender snake of smoke. Small fires were the focal points of family gatherings in the evenings. They provided light and warmth when the temperature dropped after dusk. The people woke up with the rising sun, when it was still cool, and started their fires again. As the days warmed, the fires were allowed to die out. Before sitting, Estelle and I both shook hands with the two other men, who smiled warmly and said, "Bonjour." Estelle explained our mission—that we were looking for a site to set up a new chimpanzee sanctuary. Like rural people in other places we had visited, Chief Gaspard and Colbert agreed that they would be happy to have a chimpanzee sanctuary in their forest. For them, it meant the possibility of development, of income in their village.

"Are there wild chimpanzees living here?" Estelle asked them.

"Yes!" They all answered at once. "A lot."

"What about gorillas?"

"Yes! There are gorillas, too."

I knew they thought we would be happy to hear chimpanzees and gorillas lived here, so I wasn't sure whether to believe them, but Liboz also had told us that chimpanzees and gorillas still survived in the Mbargue Forest, despite intense hunting in the area.

"How many people live in this village?" Estelle translated my question.

"Ten," Chief Gaspard answered immediately.

"How many women and children?" I knew from my previous exposure to the small village culture that his number probably didn't include women and children.

For several minutes, the four men spoke unhurriedly among themselves in their dialect. Their language had a lot of "B" sounds, which they pronounced emphatically. In years to come, the hard, forceful syllables of Bamvéle, the main dialect of villages on this side of the Mbargue Forest, would give me the impression that the speakers were angry, until a smile or laugh would put me at ease. At this moment, the men were trying to count the women and children who lived in the village, but they were perhaps disagreeing on whether to classify some as residents or visitors. Men in Cameroon can legally marry up to four women, but in the villages they often took in women they called wives without having a formal ceremony.

Estelle smoked while we waited more or less patiently. I was beginning to wonder whether I should have asked such a complicated question when, suddenly, they seemed to agree that it was not a thing they could know. "I don't know," Chief Gaspard shrugged unapologetically, not even venturing a guess. The topic was finished for them. Estelle and I exchanged confused glances. It was many seemingly nonsensical social exchanges like this one that led us to joke, between ourselves, that Cameroon was a logic-free zone. We adopted the phrase from Peace Corps volunteers we had met in Yaoundé.

"Where are the rest of the men, and the women and children?" Estelle moved on.

"Some of the men have traveled. The women are working in the farms and have the small children with them. The older children are in school." Colbert answered this time.

After this bit of small talk, we determined that Colbert could be our guide on a trek into the forest. He welcomed the opportunity to make the 2,000 Central African francs (about $3.50) we would pay him. This was well over Cameroon's minimum daily wage for people in rural areas. We left the Pajero in the village and set out walking along the road. Wearing hiking boots, with backpacks of water and snacks, Estelle and I followed Colbert, who wore flip-flops worn very thin at the heels and held his machete like a comfortable appendage near his chest. When he led us onto a footpath into the forest, the noticeable drop in ambient temperature, the receding sunlight, and the symphony of insect song under the tall forest canopy added to my pleasant sense of otherworldliness. Colbert easily outpaced us, and Estelle asked him twice to slow down. It was difficult to keep his pace and appreciate the forest at the same time. At one point he stopped suddenly with some information. "One of the hunters in our village shot a gorilla right here in this spot last year," he boasted matter-of-factly, clearly proud that his village was the home of such prowess. He didn't yet understand where we would be coming from on the issue.

"How many people in the village have guns?" Estelle asked him.

"None." He shook his head as he answered, without hesitation or explanation.

I was still registering the inconsistency, wondering how someone shot a gorilla without a gun, when he started walking again.

I wondered if he had lied about one or the other of his statements. Another option, which I wasn't knowledgeable enough to consider and which was probably the truth, was that a bushmeat dealer had placed a command and provided a gun temporarily to an experienced village hunter who did not himself own a functional gun.

When we returned to the village in late afternoon, it was much more populated. Some women came to greet us, smiling as they shook our hands. They wore traditional wraparound skirts, pieces of cloth called pagnes, the swirling blues, oranges, yellows, and greens of which, despite years of wear and harsh soap, lent some faded color to the drab surroundings of the village. Chief Gaspard approached and introduced a pretty woman standing among the others as his wife, Christine.

I noticed some young children peeping at us from the corner of a house, smiling and giggling, staying mostly hidden. One boy of about six years came out and sauntered bravely toward us, head high and arms swinging. I turned to face him with my big reassuring smile, from which he turned abruptly and ran screaming back to the cover of the house where his cowering comrades squealed and giggled in excitement.

"They've never seen white women before," Chief Gaspard explained. "Only white loggers."

He told us we could stay the night in a room of one of his houses, and we gratefully accepted. We had been prepared to set up a tent, but we were tired and happy to be spared the exertion. He issued a command in the dialect and two teenage boys appeared, each carrying a narrow bamboo cot, and indicated for us to follow them. We followed the boys and beds into a small windowless room, the only light entering behind us through the door. The boys left, and we laid out our sleeping bags on the hard cots and unpacked candles, flashlights, mosquito spray, toilet paper, soap, towels, and wine to share in the village. We had mosquito nets but couldn't see a way to easily hang them in the mud house, so we didn't use them. We were both on malaria prevention at that point so we weren't too worried.

As we were about to go out, we peeked out the door and realized that the people were eating dinner. I would learn later that the people of Bikol ate only once a day. Six men and older boys, under the roof where we had joined some of them for conversation earlier, huddled around a communal pot of goocy green stew—this mix of cassava leaves and peanuts would later become one of our staples—while eight women and several young children circled another pot several houses away. Each person had a heaping glob of "couscous de manioc," a doughy white mix of cassava flower and water. They used their fingers to pinch off bite-size morsels of the couscous to dip into the pots of stew. Hovering and patient fingers waited their turns at the pots, and between mouthfuls people chatted happily. Mothers pinched smaller morsels from their plates for young children.

"The family dinner, village style." I said to Estelle. I was surprised at how the men and women ate separately in this village. For us to join either group now would have been an awkward intrusion into an affair that impressed me for its sweet casual intimacy. Village people generally shared everything among themselves and were typically generous to outsiders, too. The people of Bikol didn't make an effort to offer us food, and I felt sure they thought it wouldn't be good enough for us. It was certainly true that we would not have eaten it from the communal pots. After a nasty case of dysentery a few months earlier, I had tried to avoid eating in villages. Estelle and I agreed to wait in the house until they were finished eating. We lit a candle and filled up on olives, bread, and peanuts we had brought with us from Yaoundé.

When we were sure the dinner ritual was over, we went out to join the men's group, carrying with us six liters of Casanova boxed wine. It was the men with whom we needed to speak. The women held no power, and we didn't even consider approaching them first. Estelle handed all the boxes to Chief Gaspard, who was clearly pleased and gestured for us to sit on a bench opposite him. As two teenage boys scooted down to make room for us, Chief Gaspard shouted a command to Christine in their local language. Momentarily she appeared, leading several other women. They carried three glasses and plastic cups of various shapes and sizes.

Chief Gaspard handed a box of wine to Christine for her to serve. She first filled the three glasses and passed one to Chief Gaspard and one each to Estelle and me, before filling the plastic cups for the other people. She and the other women shared two or three cups between them. As night approached and the temperature cooled, one of the women started a fire. A chorus of insects and animals from the surrounding forest, which were much louder at night, provided an enchanting acoustical accompaniment for our little meeting by the fire. The cheap boxed wine became more tolerable with every sip.

Estelle explained that we were a very small project with a small budget, not at all like Coron, the logging company. If we decided to locate the sanctuary here, we would bring some jobs, buy fruits and vegetables for the chimpanzees from the village, contribute to the economy in this way, but our organization would not be bringing a lot of money. We didn't want to give them false expectations. I didn't need her to translate what she was saying, because we had discussed what she would say.

Chief Gaspard and Colbert nodded a lot as Estelle spoke. When the dialogue turned more personal, I made Estelle translate every word.

One of the women asked if we had children. It was considered more respectable to be married, especially at my age, and the village people gave us the courtesy of first assuming we were married or had been. Many unmarried women in Cameroon have children, but the state of matrimony confers a higher status.

"I have a son, named Nicholas," Estelle answered. "He's with my husband in Yaoundé." Estelle had married Dana when his son, Nick, was five. Since then she and Dana had raised him, living much of the time in Africa.

When eyes turned to me, I shook my head. "No, I don't have children."

Everyone looked sad, almost embarrassed, about my plight of childlessness. There was a moment of silence as everyone looked at the ground in sympathy. It was a momentary conversation stopper.

"How many children do you have?" Estelle finally asked Chief Gaspard, a little too cheerfully.

"Twelve, with nine living."

Estelle and I nodded. It was another conversation stopper. I didn't know whether to congratulate him on the nine who lived or give condolences for the three who died, but I knew he must have suffered. When Chief Gaspard yawned, we excused ourselves and said good night.

Later I would learn that two of the other women around the fire were also Chief Gaspard's wives, and they were the mothers of some of his children. Christine was his youngest and favorite wife at that time in 1999. High-ranking men in the villages usually had many children with multiple women. Children were evidence of virility and power. For a man who had little property, his progeny could be the only evidence. In this particularly impoverished area of Cameroon, people rarely had legal marriage ceremonies. Some had traditional ceremonies in the village, while others just moved in together and called each other husband and wife. If they had children while they were together, and the man acknowledged and managed to support the woman and her kids to some extent, he would be considered the husband and the woman his wife, even if he moved on to another woman. Chief Gaspard had a mud house where he and Christine were staying, and his two older wives had houses nearby.

Back in our mud house I lay awake on the bamboo cot listening to the animal sounds of the village night. Animals that I imagined were rats, and now know were just as likely to have been geckos or huge cockroaches, crawled noisily across the raffia roof above us. Eventually, thanks to my exhaustion and a fair share of cheap wine, I dozed off.

We awoke at dawn and packed our belongings. When we emerged from the hut, a group of women carrying machetes and balancing empty baskets on their heads, many with babies on their backs, were leaving the village to work their farms—plots of land in the national forest that they had claimed, cleared, and planted. Three young children followed closely behind them. The bellies of the children were distended from protein deficiency, and even from a distance, in the bright morning sunlight I could see bald patches on their small heads from fungal "ringworm" infections. We waved to them, and the women and children smiled and waved back. One of the younger women called "Bye-bye" to me, maybe the only English she knew. I replied "bye-bye."

As we approached Chief Gaspard and Colbert, again sitting on the bench where we first encountered them, they rose to bid us farewell. With a smile, Chief Gaspard accepted the money I handed him for the room, the same amount I paid Colbert for his services as a guide. As we turned to walk toward the Pajero, Estelle assured the chief that we would see them again.

We drove about a half a mile from the village and parked along the dirt road where a small trail cut into the forest. Overconfident and eager to get a feel for the forest at our own pace, we set out on foot, equipped with my compass and a machete. We followed the tiny trail, which intersected other tiny trails. I thought the paths must have been used by village people for hunting and for gathering things they needed. The great variety of trees, undergrowth, colorful insects, and strange sounds in the African forest were fascinating, exotic. The Mbargue Forest was drier than some we had explored. It was actually a mosaic of savanna and forest, with the forest especially dense along the routes of small streams. Natural habitats for chimpanzees include both rain forests and drier forests like this one.

Estelle and I walked happily for over an hour, sticking to the trails. When we started trying to make our way back to the dirt road where we had parked the Pajero, we were confused about which trail to take. We knew we needed to walk northwest to get to the road and our compass pointed the way, but none of the winding trails were going our way. We would need to use our machete to cut a straight track back to the road. As I gazed upward, trying to see the direction of the sun through a combination of dense canopy and cloud cover, unnecessarily double-checking the accuracy of our compass, I heard Estelle scream something unintelligible. By the time I turned and saw her retreating backside, she was barely visible on the narrow trail, running fast away from me. All at once, dozens of painful bites on my legs, and in my underwear, added misery to my confusion.

"What the *hell!*" I shouted to no one in particular, since Estelle was long gone.

At exactly the same moment, my eyes focused on the sea of insects teeming across my shoes and my ears finally registered what Estelle had said: *"Ants!!"* I squealed and cussed and bolted after her, trying to reach a part of the trail without ants. Looking down at the trail, I didn't notice as I rounded a bend that Estelle was standing right in the middle of the path, practically naked, picking off ants. I crashed into her, sending her stumbling forward. She caught herself against a tree trunk and cushioned my fall as I clutched at her and stayed on my feet.

61

Estelle was not the type to suffer this clumsy slapstick silently, but it was no time for useless scolding. As soon as she was solidly on her feet again, she turned her attention to the ants still biting her.

Within seconds I too was mostly naked and picking at little clinging flesh eaters, which refused to be flicked off easily. I soon learned to squash them quickly between my fingernails, so they couldn't cling to a finger and bite again. Like virtually everyone else working in Central African forests, I came to know these carnivorous ants well. They don't sting like the fire ants of Mississippi that I knew as a child, but they attack by the millions and literally eat their victims. They can quickly kill an injured or trapped animal. For healthy humans walking in the forest, they are merely a painful annoyance, but people who are accustomed to the forests don't take their eyes from the dirt for long.

We finally managed to wander out of the Mbargue Forest onto the road and find the Pajero by four o'clock. We headed toward the small town of Minta, where we planned to spend the night. Our map showed us that the road was unpaved all the way—the kind of narrow red dirt track that connected towns and villages in much of rural Cameroon—and we knew it was too late to be starting out. We would be finishing the drive at night, which could be dangerous. Bandits could stop a vehicle on a rural road by laying down a long two-by-four with nails sticking up from it. Cars that didn't stop couldn't go far on four flat tires. Police also used the nails-across-the-road technique to stop cars at checkpoints, which sometimes seemed impromptu and in the middle of nowhere, so it could be hard to tell who was trying to stop you. It was best to avoid traveling at night. When we were two hours into the drive and the clouds darkened, I suspected that our misfortunes of the day were not quite over. As twilight descended, so did a torrential rain, and the combination limited my visibility to only a couple of yards. Within a few minutes the road was slick and the Pajero's four-wheel drive didn't seem to be working at all. I managed to slip and slide slowly forward until I was forced to stop at a pond of water that stretched across the road.

I had seen rain-filled basins making travel difficult on Cameroon's roads. During rainy season, heavy logging trucks made deep basins in the soft mud roadways, and rain accumulated in them. But this basin of water, extending two yards beyond the road on both sides, was a small pond that completely blocked the road. The road was too narrow and slippery to turn around. I tried to back up, but immediately realized that backward driving afforded me even less control than I had going forward. I was sure to slide off the road and get us stuck if I didn't stop. I turned off the car. The rain wasn't letting up, and it was getting dark quickly. We didn't know what lay ahead of us, on the other side of the pond, but the last village we had passed, where people might have welcomed us, lay many miles behind us.

"We have to stay here in the car until morning," I announced to Estelle.

"No way!" Estelle exclaimed. She was one of the bravest, most capable people I knew, but she had a little quirk about sleeping in the dark. She needed light to sleep. I knew that was going to be problematic. There was no light at all from the sky. No way our flashlight batteries would last until morning. I felt badly for Estelle, but I secretly thanked the invisible stars that she had used the last of our candles in the village the previous night—at least I wouldn't worry about the car catching fire, as I often had in hotel rooms when she left candles burning.

"There's nothing else to do," I stated. "When it's light, we'll figure out how to get out of here. We'll be safe. With all this rain, no one else will be able to come this way in a car, and no one will walk here tonight." Since I had been a child I had always reassured people compulsively. One Saturday evening as a teenager in Mississippi, I was driving on Interstate 55 with my cousin Denise in the passenger seat when a drunk driver hit us hard from behind. Completely out of control, my mother's car completed two 360-degree turns and crossed a wide median before finally coming to a stop facing oncoming traffic. During the terrifying spin, as Denise screamed, I struggled to raise my voice above hers to reassure her over and over, "We're okay! We're okay!" Now, stranded on this dirt track in Central Africa, I told Estelle, perhaps a little too casually, "We'll be fine!"

We were out of drinking water and already thirsty, so I used my pocketknife to cut two plastic water bottles in half and put them on top of the Pajero to catch some rain. Estelle crawled onto the backseat and through the long, dark night we talked—mostly adding up the pros and cons of the Mbargue Forest as our new sanctuary site.

"The forest itself is beautiful, the villagers are nice, and once the bridge across the Yong River is finished, the train going to Bélabo will give access year-round," I said, cheerfully pointing out what Estelle already knew. Jean Liboz was building a bridge across the river to facilitate movement of Coron's logs. When it was finished, the town of Bélabo, which had train service from Yaoundé, would be only fifteen miles from Bikol. Two of the other potential sanctuary sites we had looked at were inaccessible during rainy season, which ruled them out.

"There are enough villages to grow food," Estelle said. "Farming is part of their culture, which is really good." Down south in Campo, another place we had explored, the people lived off the sea and did little farming.

"There aren't enough people in Bikol to hire for all the positions we'll eventually need to fill," I said.

"But it'll be better to hire from different villages," Estelle said. "For one thing, when someone in the village dies, we won't lose the whole workforce for the funeral." I laughed, but Estelle wasn't kidding. If I had known then what I did a few years later, when I had seen how many funerals there were in the village—how often young people who shouldn't have died in the twenty-first century did, indeed, die—I wouldn't have laughed.

We agreed that the Mbargue Forest met our criteria better than any other place we had explored. Our conversation, as usual, touched on personal issues too—difficult childhoods, families, lost loves. Estelle was a strong and capable woman, wise beyond her twenty-six years in many ways, but she was fourteen years younger than I was. When we spoke of personal relationships, when her emotional vulnerability showed through her tough-girl façade, I often felt maternal toward her. I never told her this though, because I thought she would have found it condescending.

At one point, after dozing off, I woke to the suffocating stink of heavy cigarette smoke. I could see the glow of Estelle's cigarette. The rain had stopped, but it was still black outside—no trace of moonlight or starlight filtered through the persisting clouds. I rolled down my window and breathed in the cool, clean air.

"You're giving me cancer," I said grumpily.

"Sorry." She took another long drag on her cigarette and rolled down her own window a little more.

At dawn, the pond looked even bigger than it had the night before. The road was still too slippery to back up. We had two options: wait at least half a day for the sun to dry the road a little, then try to back up until we could find a place to turn around, or try to drive through the pond. I took off my shoes, rolled up my pants, and walked out into the water, soft mud squishing through my toes. The water was up to the middle of my thighs at its deepest part.

"It's definitely too deep to drive through," I said.

"Maybe we can drain it," Estelle said hopefully.

By the time I had waded out of the water, she had brought out the machete, our only tool that could be used for digging, from the back of the Pajero. This was the beginning of my understanding to always carry a shovel when I traveled on bush roads. Estelle strode purposefully toward the banked mud that formed the wall of the pond off the side of the road and started hacking at the mud with the machete, slinging reddish-brown globs everywhere. Soon she had made a shallow V-shaped trough across the top of the bank, finding that the deeper she cut, the more solid was the mud. By removing chunks of it with the blade of the machete, she was trying to make a trench in the bank of the pond deep enough for the water to drain off to the side.

"You're brilliant!" I encouraged her. By this time her clothes, face, and hair were dappled with mud.

It was slow going and after about five minutes of swinging the machete, Estelle was exhausted. I took over until my own arms stopped cooperating and she took the machete back. Before I got another turn with it, the sludgy brown water was flowing slowly through our trench. We deepened the trench several times when the water flow slowed to a trickle. Then we labored another half hour to make a second trench on the other side and drained more water there.

When I waded out again, the water was just below my knees.

"I think we should go for it," I said, and Estelle nodded her agreement. If we got stuck, we might be in real trouble. We were out of drinking water, and I figured we were at least fifty miles from Minta.

I cranked the Pajero, we prayed silently, and I started through the water, slow and steady. Where the water was deepest the wheels started spinning and the Pajero fishtailed, creating waves that lapped the walls of the mud banks. For a second I thought we would get stuck, but I kept my foot lightly on the accelerator, and we kept inching forward until the wheels found traction again. After what seemed like much longer than the few seconds that it was, we arrived on the other side.

"Yeah!!" We both cheered loudly with relief.

"Good driving!" Estelle complimented me.

By 10:30 A.M. we had stopped at an auberge (small hotel) in the town of Minta. There was no running water, but we paid the attendant to bring us each a bucket of water from the well to clean the dried mud from our hair and bodies. We had decided to detour through Minta on our way to Yaoundé in hopes of meeting with the divisional officer (D.O.) for this district in which the Mbargue Forest stood. It was an important step. He was one of the local representatives of Cameroon's many-tiered national government, and we would need his support if we were to build a sanctuary in the district.

Minta had no phone lines, so we took a chance that the D.O. would be there, and we got lucky. We found Mr. Ndang Ndang Albert, a stocky, slightly balding man in his midthirties, in his home just behind his office. His house was made of plastered concrete blocks and was cool and comfortable inside. In the living room, where he received us, the maroon-colored concrete floor was partially covered by a large, pretty Asian-style rug, which was a subdued hue of dark blue with swirls of small multicolored flowers. He invited us to sit on an upholstered brown couch, while he sat kitty-corner to us on a chair covered with the same fabric. I imagined the effort he, or someone, had expended to get the furniture and rug to Minta. He offered us Cokes, which were room temperature and still tasted like nectar of the gods to us. In a meeting that lasted no more than half an hour, he promised that he would welcome our sanctuary as a form of development in his impoverished rural district.

Happy with our accomplishment, Estelle and I headed toward Yaoundé under clear, blue skies. We had found our sanctuary site.

six

SHACKLED

En route from the Mbargue Forest, we decided to make a stop at Luna Park Hotel in the town of Obala, just an hour outside of Yaoundé. Karl Ammann had told us that two adult female chimpanzees were held on chains there. It was shortly before dusk when we crept up the long dirt driveway of the hotel in the 1990 blue Pajero, which in spite of its age would have been shiny and pretty without the thick coating of dried red-brown mud that clung to the lower half, and the same-colored dust that coated the rest. We stared out in wide-eyed anticipation through the half circles that had been cleaned by recently replaced windshield wipers and most of a reservoir of wiper fluid.

"Can you see them?" I asked Estelle. As I drove toward the cluster of buildings at the end of the driveway, Estelle scanned the landscaped lawn for two chimpanzees.

"There's a big baboon on a chain," Estelle answered. Then, moments later, "There's one of the chimpanzees! A big adult." Estelle looked to the right through a place she had cleaned on her window. "And there's another smaller chimp behind her."

As I steered the slow-rolling Pajero, I took a quick look. The two chimpanzees, chained in the shadows, were separated from each other by an expanse of grass. They were much too far apart to touch each other.

Fifty yards ahead I parked the Pajero in the gravel parking area to the right of the driveway, and we walked across the driveway to the hotel office to rent a room. Adjoining the office was an open-air restaurant, and on the opposite side of the restaurant was a covered sunken veranda. As we approached the office, we veered casually toward the veranda to count seven skinny juvenile monkeys of several species tied by short ropes around their waists to the vertical supporting columns of the veranda's roof. Diners in the restaurant could look down to see the monkeys, and the monkeys could see them eating. As we watched, some monkeys sat despondent and others darted about anxiously in all directions as far as their ropes would allow.

Luna Park Hotel was the only visitor attraction in the small town of Obala. We had learned that a Senegalese family owned the hotel, and that it had been a popular destination for well-to-do Cameroonians for decades. On weekends, wealthy politicians and businessmen brought wives and children to have lunch, fish in the small river that ran through the center of the vast, manicured hotel grounds, and laugh at the monkeys. Most people didn't distinguish between monkeys and chimpanzees, although the contrast was stark. As tailless great apes, chimpanzees are much closer to humans than they are to monkeys.

We rented a room and found our number on a door in the long, narrow, single-story building. The room was starkly furnished with two single beds, two wooden bedside tables, and a simple wooden wardrobe for clothes. The baby-blue walls were bare of anything but a few dirt smudges, and the only light came from a single bulb in the center of the ceiling. The bedding looked clean, and the bathroom had cold running water and a flushing toilet. I hated cold showers, but any running water seemed a luxury, and it was enough to satisfy us. We dropped our backpacks on the beds and rushed back out, eager to see the chimpanzees.

We hurried back to the Pajero to pick up two big plastic shopping bags of bananas and papayas, and another of unshelled peanuts, all of which we had bought from several small villages en route from Minta. With our heavy loads of fruit and peanuts distributed between us, we walked back down the driveway of the Luna Park Hotel to meet the two chimpanzees before dark.

We walked toward the bigger chimpanzee, closest to the driveway, depositing the bags in the grass before going as close as we dared. The chimpanzee was overweight and sat almost motionless within the barren circle of dirt, where no grass would grow, at the base of a big moabi tree. In the dusk, it was hard to see her dark face clearly, and my eyes were drawn to the heavy-gauge chain with its oversize padlock shining in the twilight against the black hair of her neck. The chain led from her neck to the trunk of the tree, the end looped around the tree and linked back to itself with another padlock.

I tried to guess the length of the chain from the coil on the dirt, wanting to get as close as possible without going within her reach. We didn't know her temperament.

"I'm guessing about three yards of chain," I said to Estelle. She nodded and we walked a few steps closer before we squatted about four yards from the chimpanzee. We could confirm that she was female, and we could make out the features of her face. Her fully alert brown eyes met mine and held them. Her long black face was beautiful.

"Hey, girl. We're here to help you," I called to her.

Estelle panted softly in greeting, and the chimp's gaze shifted to her. She was curious, but she didn't move. She didn't seem to expect much good to come her way from us, but after a few moments she looked away from Estelle's face and cocked her head to look directly at the plastic bags in the grass two yards behind where we now squatted.

"What's in the bags?" the chimpanzee seemed to ask. Her meaning was obvious.

Estelle walked back to retrieve a bag of bananas. When she pulled out several, the chimpanzee moved quickly toward us, causing us to jump back reflexively, although we were already out of her reach. Standing upright on her two legs, she looked even bigger than before. She grimaced in excitement, baring her teeth, as her high-pitched, pleading vocalizations stuck wet and garbled in her throat, almost like she was choking. She reached out both arms toward the bananas, but before Estelle had time to give them to her, she pulled her hands back in close to her chest and began flopping them rapidly up and down. Chimpanzees can be quite melodramatic in their expressions of need and frustration, but there was a quality of hysteria and desperation in this chimpanzee that was heart wrenching and scary at the same time. It spoke of needs unmet for far too long. Estelle tossed her the first banana, which she caught easily in her right hand. Immediately silent, she held out her left hand to catch another that Estelle threw. It was a practiced skill. As she sat down to peel the first banana quite delicately and deliberately, in stark contrast to her urgent, frenetic appeal just moments before, her sweet grunts of contentment relaxed and delighted Estelle and me. When she was eating her second banana, Estelle walked closer, stretching out her arm to offer her two more. Without getting up, momentarily holding the banana she was eating between her lips, she used both her hands to gently accept the bananas from Estelle's hand.

From the time we arrived with the bags of fruit, I was aware of the other, smaller chimpanzee moving left and right, back and forth close to the brick wall that bordered the hotel, about twenty yards from where we were standing. I noticed that a few pieces of zinc roofing material extended from the brick wall over her head to provide some minimal shelter from the sun and rain. As I approached the chimpanzee with four bananas connected by a single stalk, she stretched her shackle to its limit, holding out one begging hand toward me, while the fingers of her other hand encircled the loop of chain around her neck to relieve the pressure. The chain, only about four feet long, tethered her to a two-foot-diameter concrete slab, level with the dirt in which it was anchored. Staying out of her reach, I held out the bananas. She snatched them from my hand and retreated slightly to gain some slack in the chain before sitting and peeling one. While she ate, I walked a little closer and squatted to get my head on a level with hers and get a better look. I saw that this chimpanzee was an adult, too, but she was tiny compared to the first—small of stature and painfully thin, with grayish hair. After she devoured the first two bananas, she paused for a second or two to look at me. Her bottom lip hung open, revealing pale gums, and brown, stained teeth. It was clear that she was anemic and malnourished. The light was dim, but I could see what appeared to be diarrhea smeared across the concrete and in the surrounding wet dirt. There was no food debris, which made me suspect she was given very little to eat.

Walking back to the bags for more fruit, I saw the big chimp politely take half of a big papaya from Estelle's hand and place it neatly beside the small banana pile in front of her—grunting and smacking happily as she peeled and ate another banana.

I took four more bananas and the other half of the papaya and returned to the small girl. She sat with her knees bent, her feet pidgin-toed and flat on the ground, chewing on the last of the four bananas. When I approached her with the additional fruit, she placed a protective hand over the banana peels she had collected in a pile close to her outer thigh. She must have been saving them for later. Without getting up, she reached out her bony arm three times in quick succession to twice take two bananas and then the papaya as I held them out for her. She placed the bananas in the protected small space between her feet while she started on the papaya—my first indication that papaya was her favorite.

Finally, Estelle and I put neat piles of peanuts within reach of each chimp, careful not to scatter them, since the rapidly approaching darkness would make foraging difficult.

We had saved bananas, a papaya, and some peanuts for all the monkeys. Estelle took food to the skinny adult baboon, whom we could see pacing frantically on his chain closer to the driveway entrance, and I took some back to all the small monkeys tied on the veranda.

The monkeys, who were all too thin, snatched the fruit I held out for them as though I might change my mind about giving it to them. As they gobbled the fruit, they used their small hands and feet to protect the peanuts I placed in front of them. I so wished we could get these monkeys out of here, too, but I knew that none of their species were considered to be in danger of extinction and that laws protecting them were much more lax. Although there had been little enforcement, it was illegal under Cameroon law for private individuals or businesses to keep endangered species, including chimpanzees, in captivity. On the other hand, it was legal for them to keep these monkeys as long as they bought permits. As sad as the plights of these monkeys were, we were building a chimpanzee sanctuary, and we would focus on rescuing the chimpanzees. At least for now.

After all the food was distributed, I hurried into the bar of the hotel to buy two bottles of water, because the chimps didn't have water containers near them and it was too dark for us to search about for any. After pouring a small amount of water into a chewed-up plastic bowl Estelle had found for the adult baboon, and a small amount into seven bowls I found scattered around the veranda for the small monkeys, we gave one half-full bottle to each chimpanzee. They both lifted the bottles and drank eagerly.

By this time, it was too dark to see without our flashlights. Dusk in this part of the world is short, and night comes quickly. Cameroon is located just north of the equator, which means we have close to twelve hours and thirty minutes of light every day, with little variation. Estelle and I were exhausted and hungry. I sensed, rather than actually saw, the chimpanzees watching us as we walked along the dimly lit driveway toward the hotel restaurant. For the first of many times to come, I felt an aching pang of guilt in my freedom to walk away and leave them.

At the restaurant Estelle and I split a big, slightly chilled bottle of 33 Export, one of the heavily advertised beers brewed by a company owned in part by Cameroon's president, and ate spaghetti omelets and French fries. We learned from the waiter that the bigger chimpanzee was Dorothy, and the small one Nama. He told us the manager of the hotel, who was one of the sons of the owner, would be there the next morning.

We ate fast and went to our room. Estelle braved a cold shower, but I decided to skip it and go straight to bed, despite feeling dirty from the long drive. I was exhausted. Being prone, even on the worn, lumpy foam mattress, was a relief, although sleep didn't come to me until hours later. I obsessed about Dorothy and Nama sleeping on the hard dirt outside. Would we manage to take them away from here? We hadn't even built the sanctuary yet, and I didn't know how long it would take. Could we move quickly enough to save Nama, who was obviously sick? Could I help her with medical treatment even before we would be able to move her? Would this Senegalese family that was actively collecting primates agree to let us take the chimpanzees from here? What kind of struggle could we face with these hotel owners? Would the Ministry of the Environment and Forestry be willing and able to stand up to them? The patriarch of the family had ties to Cameroon's former head of state and was considered politically powerful. However, I thought with a glimmer of optimism, the former head of state died in exile after the current president took power. I wondered if the family was connected in any way to the present administration. On and on my thoughts ran, until sometime after two o'clock I finally fell asleep.

Estelle woke me early. We dressed quickly, grabbed bottles of Coke from the restaurant because we didn't want to take the time for coffee—we knew the restaurant wouldn't have paper to-go cups—and said a quick hello to Dorothy and Nama before rushing to the market in Obala to buy food for them. Someone had given them each a pile of palm nuts, which are high in saturated fat and therefore not good for them. Dorothy's pile was much bigger than Nama's, but neither chimp ate them with much relish.

Obala's open-air market was about three-quarters of a mile from the turnoff to the Luna Park Hotel driveway and only a block from the paved road that led to Yaoundé. Its façade was a row of ten gray ancient-looking wooden stalls perched side by side on packed brown dirt. Each was crammed with a variety of colorful fresh produce, hawked by cheerful, mostly female vendors. Behind the first row of stalls were several additional rows of tables, made of the same worn gray wood, but without walls. A patched, rusting corrugated zinc roof hung over the tables, with gaping open spaces that let in light. Estelle haggled in French over the prices of fruit with the women along the first row, and since my presence was providing no benefit to the process, I wandered through the back alleys of the market.

The back rows had less of the fresh fruits and vegetables and more dried foods—big metal basins of corn for grinding into flour, shelled peanuts (called groundnuts in Cameroon), red beans, white beans, soybeans, and melon seeds. Table after table had the same things. Laid out beside the basins on some of the tables was an assortment of dried spices I hadn't seen outside of

Africa and didn't have any idea how to use. Most of the women spoke to me as I browsed their tables, trying to convince me I needed to buy one thing or another. "No, merci. Je ne parle pas français." *No, thank you. I don't speak French.* Accompanied by a smile, it was my standard response. However, I had practiced this one phrase enough that it sounded like I really did know how to speak a little French, so people kept talking to me. "Je ne comprende pas!" *I don't understand!* It was my other practiced phrase.

The last row of Obala's market was dedicated to meat. Several tables were covered with the smelly fly-covered flesh and bone of domestic animals. A table on the end had stacks of smoked bushmeat—blackened body parts of various forest animals, many unrecognizable. Perusing the table, I identified several limbs of small monkeys, but nothing that looked like chimpanzee or gorilla parts. On the dirt beside this table was the freshly killed bushmeat. Laid out on their sides were two putty-nosed monkeys—sweet-looking little guenons with gray-black hair and brightly contrasting white noses—who had been shot and probably died quickly. Beside the monkeys was the body of a small dark-gray antelope called a blue duiker, who had suffered long—one delicate leg, broken and twisted, told the story of his panicked, futile thrashing in a trap. And beside the duiker was the freshest bushmeat of all. A juvenile dwarf crocodile, slightly more than a yard long with his short front legs tied tightly behind his back, was still breathing. His legs were already swollen and purplish from lack of circulation. As I squatted beside him, he opened his eyes slightly wider, and I was careful not to get close to his mouth. I wanted to help him, but how? Even if I could free him, he would lose his legs. He could not survive now. I could buy him and show him the mercy of a quick death, but I didn't really know how to kill a crocodile. I remembered the machete in the truck, but I wasn't even sure where to hit him. The underside of the neck would be best, I supposed, but how many strikes would it take? I surely didn't trust my aim with a machete, and while my intention would be to end the crocodile's suffering, I might end up causing him more. Also, I knew very well that it wasn't wise to reinforce the hunting trade by purchasing bushmeat. Stand up and walk away, I willed myself. In the United States, most animal cruelty was hidden from public view behind locked doors. In this country, where it was overt and ubiquitous, I needed discipline and a more callous heart to stay focused on my mission. Wishing him a fast death and reassuring myself that his brain was very small, I left the crocodile to whatever his fate would be.

Since I had taken the keys with me, Estelle was forced to wait for me outside the Pajero. "What the hell were you doing? It's hot out here," she complained loudly and legitimately. She couldn't come find me, carrying all the heavy fruit she had bought.

"I'm sorry," I said, and I was. Sorry for keeping her waiting in the heat. Sorry for leaving the crocodile to a slow death, choosing the easy route for myself.

"What were you doing?" Estelle repeated.

"Nothing," I said. "Just looking at the market. What did you buy?"

Estelle had bought bananas, papaya, mangoes (the first of the season), corn on the cob, peanuts, and long baguettes of white bread—enough of everything for all ten primates at Luna Park. We stopped at a small supermarket to buy two plastic cups of yogurt for Dorothy and Nama. Estelle and I both had read about the diets of free-living chimpanzees, but the wild fruits and leaves they ate weren't available to us. The fruits and vegetables and peanuts that Estelle bought were staples of the chimpanzees at LWC and also were part of the diet of the chimpanzees at her sanctuary in Guinea. In addition to all kinds of fruits and vegetables, we had seen Jacky, Pepe, and Becky enjoying bread from the hotel restaurant, and yogurt was a special treat I often bought for them. We knew we had food that Dorothy and Nama would like.

Back at Luna Park, as we carried the food toward Dorothy and Nama, an attendant approached us.

"The chimpanzees are very dangerous. The big one can be vicious," he informed us. "I am the only one who can go near them. I'm the one who cleans after them and feeds them." While I waited for him to speak and Estelle to translate, I watched Dorothy and Nama. They were watching us and waiting impatiently—Dorothy sitting at the limit of her chain and Nama pacing jerkily back and forth.

"What do they normally eat?" I asked him, through Estelle.

"Palm nuts and bananas," he said. "They will also drink beer and smoke cigarettes. If you have cigarettes, I'll show you." Estelle explained to him that I was a veterinary doctor, that we worked with chimpanzees, that cigarette smoking was bad for chimpanzees, and that we did not want to see them smoke. I suggested that we could write a list of appropriate foods so he could try to get a variety of them to feed the chimpanzees and other primates every day. We showed him that we had food in our bags and informed him that we would now give it to the chimpanzees ourselves. He looked skeptical, hesitating as he formulated his objection. Without waiting for it, we turned and walked quickly with our several bags of food to a patch of manicured grass between Dorothy and Nama. Dorothy grimaced and bounced on her back feet in excitement, but her extreme anxiety of the previous day was absent. Nama, too, was calmer as she watched us remove food from the bags. Estelle doled out the fruit, while I handed each of them yogurt and bread. We approached the chimpanzees less cautiously than we had the day before. We were now acquaintances, with a positive record—albeit short—of benevolence.

While they were enjoying their food, I fed the baboon, collected the water bottles to get fresh water, and distributed some food and water to the small monkeys on the veranda. When I came back out, Estelle was sitting close to Dorothy, grooming her outstretched leg. At the same time, Dorothy gently groomed the top of Estelle's head. How sad it was that the attendant and others at the hotel had misunderstood Dorothy so completely. How many years had it been since anyone touched her in a loving way? How long since she was allowed this simple pleasure of grooming someone else? As I watched the sweet scene between Dorothy and Estelle, which I knew represented a blossoming friendship, I longed to change places with Estelle.

I turned toward Nama, who also was watching Dorothy and Estelle. As I walked slowly within her reach, she took my arm, and I allowed her to pull me in close to her. I sat down beside her in the wet dirt, trying to avoid the diarrhea. She looked at my face curiously for a few moments, glancing *at* my eyes but not really looking into them. She was inspecting me, rather than trying to communicate. After a minute or two, her hand hovered in front of my face, her lips grew taut in concentration, and her mouth smacked open and closed rhythmically. Understanding that she was about to groom me reassured and relaxed me, but then her fingers on my face were not really so gentle. She was digging at the corners of my eyes in a way I didn't enjoy. I turned my face away. When I looked back at her, she perused my face again briefly, and then tried picking my nose with a finger that smelled of feces. I turned away again. I clacked my own mouth and tried to groom *her* face, but she didn't like it either. She turned her head to escape my hand as I had done with hers. This wasn't going perfectly.

Finally, when I lowered my hands to groom her chest, she pushed her shoulders back and straightened her neck to give me good access. I moved both my hands over her chest the way I thought another chimpanzee would—parting the grayish hairs, flicking off dirt particles, gently scratching at blemishes on skin stretched tautly over easily discernible ribs. After about ten minutes, Nama lowered her head and returned her chest and shoulders to normal posture. When I looked up to see what she wanted to do next—not more face grooming, I hoped—her eyes were seeking mine with a desire to communicate that startled me. While she held my gaze, she took my right hand and placed it purposefully on the chain around her bony neck, rubbed bare of hair by the shackle. Her lower lip hung open, and her eyes were steady, beseeching. She was requesting the freedom she needed most of all and was expecting no less than simple action as an answer from a friend. All I could give her was a promise that she couldn't understand, although I meant it with all my heart.

"Nama, I *will* take that chain off you, just as soon as I am able to. I will never rest a single day until I do."

Suddenly, Estelle's voice surprised me in its proximity to us. She had approached without me noticing her. She was saying that a group of three men, including the attendant, were watching us.

"One of them might be the manager. We should go talk to them," Estelle said.

I agreed that I should accompany Estelle to speak with the men. I had become totally relaxed with Nama, but when I tried to stand up, the emotional tenor changed abruptly. Nama grabbed my arm tightly and pulled me back down, causing me to fall hard on the dirt. While I was thinking about how to handle the situation, she took my hand and stood upright on her two legs. When I managed to get on my feet, she started walking me in tight circles around the concrete slab to which the end of her chain was attached. She mostly walked upright, only occasionally putting down one hand on the dirt. With her other hand, she held on to my hand tightly, stubbornly. Around and around we went. She may have looked frail, but she was clearly a lot stronger than I was. I wasn't sure what she might do if I tried to pry her hand from mine or spoke harshly to her. I really wanted us to be friends, and I really *didn't* want her to bite me. The loss of control frightened me.

"Nama won't let me leave at the moment," I told Estelle, who had seen what was happening. I tried not to sound nervous about my capture.

"Do you want me to try to help you now?" Estelle asked.

"Go talk to the people first," I said. "I don't want them to get annoyed that we're ignoring them."

Since I couldn't speak French, we had already planned what Estelle would say, which was basically just the truth. We were starting a sanctuary in the forest for adult chimpanzees, and we wanted to take Dorothy and Nama there so they could have a better life. She would also try to find out how long Dorothy and Nama had been at the hotel.

Nama quit walking me in circles, and we sat back down. She released my hand, and once, I tried to scoot away. I hoped that by staying on her level, instead of standing up abruptly, I would be able to slip out of her reach, but when the space between us increased to a few inches, she grabbed my arm tightly. *Stay with me!* Minutes earlier Nama had owned my heart, and now I hated being subjugated by her. She would no longer meet my gaze or even look at my face, but she could not let me go and face being alone again. Suddenly, my empathy for her smothered every other emotion. If I so detested the loss of my free will for even a few minutes, how unimaginably horrible it must have been for Nama, and for Dorothy, to be so cruelly deprived of even the slightest autonomy and choice. I scooted closer to her to entwine my arm through hers. She could have her way, in *this* at least. I would stay as long as I could.

Sitting arm in arm with Nama, I watched Estelle talking to the three men. Soon, she and one of the men strolled away from the rest. He towered over Estelle and appeared comfortable as they talked. He was casually dressed in Western clothes, but he carried himself with the poise of Cameroon's affluent class.

Estelle gestured toward me. I waved stupidly, and he nodded. In a few minutes, Estelle and the man shook hands, and he walked back toward the office of the hotel, the other two following.

"How did it go? Was that the managing son?" I asked her when she approached.

"It was. According to him, Dorothy has been here over twenty-five years, and Nama came in 1984. He said he will talk to his family about letting us take them."

"Did he seem opposed to the idea?" I asked.

"He wasn't particularly negative about it. I told him you were a veterinarian examining the chimpanzees, and we would bring back some medicine for them," Estelle said. "He seemed to like that."

Unfortunately, Estelle needed to get back to Yaoundé to attend a play that evening at her son Nick's school. As we were discussing my captivity and deciding what prize—something the restaurant had for sale—we might barter for my freedom, Nama suddenly moved away from me to the opposite side of the concrete slab. She glanced across at me one time, quickly, and then stared out past me, to the left of me. As simple as that, she told me I was free to go, and her resolve broke my heart again.

"Nama, look at me," I called to her. "I'll be coming back soon." But she had retreated into herself, done with me for the day.

I had no idea how much of a fight it might be, or how long it might take, to rescue Nama and Dorothy from Luna Park, but I intended to give Nama some medical treatment within days. I collected a sample of her feces in a plastic bag I had carried in my pocket for that purpose. I would have it checked for parasite eggs when we were back in Yaoundé. Intestinal worms could be at least partly to blame for Nama's diarrhea, anemia, and wasting away. I would run the tests and come back soon with medicine. I would spend time with Dorothy next time.

As we walked to the Pajero, I stopped and looked back at Dorothy and Nama for a minute or so—from one to the other. Their loneliness and boredom were palpable. I tried to comprehend the aching, crazy-making frustration in such extreme confinement, but I could not even imagine the depth of despair that such a life must bring. How had they maintained their sanity? A painful, burning lump of fury formed and expanded in my chest and burned my throat, seeking escape. I wanted to sob. I longed for the release that tears might bring, but my eyes were dry. Anything but effective action seemed self-indulgent. My aching fury was as trapped in me as Dorothy and Nama were trapped.

seven

THE VILLAGE

Soon after our search for a sanctuary site in Cameroon ended at the Mbargue Forest, Estelle took over operation of a chimpanzee sanctuary in the West African country of Guinea. When she and her husband, Dana, had left Guinea two years earlier, she had taken the eight baby chimpanzees in her care to a larger sanctuary in Haut Niger National Park. Now the director of that sanctuary had just resigned, and the government of Guinea asked Estelle to come back and fill the position. It was a position that would give her a great deal of autonomy in decision making, but also included raising *all* the necessary operational funds.

Estelle moved back and forth from Cameroon to Guinea. When she *was* in Cameroon, she was busy raising funds for her Center for Chimpanzee Conservation in Guinea. Her availability to help me set up the new sanctuary was limited, although she really came through as a friend and collaborator at some crucial moments.

On May 31, 1999, the minister of the Ministry of the Environment and Forestry in Yaoundé signed an official letter giving us permission to establish a chimpanzee sanctuary on government-owned land in Cameroon. Until that point, we had been moving forward with verbal assurances and a letter from a lower official in the ministry. Getting the signed letter from the minister gave us real authority from the national government to move forward.

I had just hired Kenneth Fonyoy, a twenty-eight-year-old Anglophone from the Northwest Province of Cameroon, to be my driver and translator. Kenneth was smart. He had a good sense of humor, but his handsome face often wore a solemn expression. His smile was pretty, but not quick. He wasn't a big man, but his cocky attitude made him seem larger than he was. I thought he was chauvinistic, and his fast driving got on my nerves. My monitoring of the speedometer, like a mother with a teenaged son, undoubtedly got on *his* nerves. I got the distinct impression that he thought this job of driving me around was beneath him, and I was quite sure he thought I was a little crazy.

When he threw a Coke can out the window, I insisted he stop the car and go collect it.

"We don't litter from this car!" I explained a little too emphatically, faced with the glaring evidence that everyone except me, and perhaps other people riding in this particular car, did indeed litter in Cameroon. When he picked up his Coke can, he pointed to three others on the ground within view and asked sarcastically, "Would you like me to pick up those, too?"

When he captured a stunned and frightened quail that he had accidentally, although fortunately from his perspective, hit with the Pajero, he was thanking God for the free dinner until I thwarted his plans. I insisted he place the bird on the grass, and as we watched her lying almost lifeless before us, Kenneth assured me, "She will die here in the grass, a benefit to no one." When she suddenly flew away as if nothing had happened, he rolled his eyes and walked silently to the car. For him, I was unreasonable and annoying, but he needed the job more than he wanted me to know. I needed him even more.

By the time I hired Kenneth and we received the minister's letter, we had decided to locate the sanctuary in a particular section of the Mbargue Forest near the village of Bikol. It met the essential criteria we had established, and Liboz would make our driveway, clear our camp, and transport our supplies. Although Cameroon's national government favored our selection of the Mbargue Forest, the section we had chosen was considered the traditional territory of the village of Bikol. The current national government, led by President Biya since 1983, encouraged people to "help themselves"—as in "God helps those who help themselves." This meant eking out their existences on free land in remote forests where they were free to hunt, gather, and farm. As a result of the "free land" policy and the burgeoning human population, new settlements of squatters were continuously springing up, especially along logging roads that created access. However, like hundreds of other small, isolated villages scattered throughout Cameroon's forests, Bikol had predated the existence of the Republic of Cameroon as an independent country with its current boundaries and its national government, and it preceded Liboz's logging road by four decades.

Bikol was created in 1958 when two men who were natives of the larger village of Mbinang decided to break off from it. About two and a half miles from Mbinang, the founders of Bikol found the small Ndian River, running through the bottom of a steep ravine, on its way to the Sanaga River. They perched their two houses on level land about a half mile up an incline from the little river, which would be their water source. The name Bikol means "the King" in their Bamvéle dialect. The two men brought relatives to join them in Bikol, and with multiple wives, they swelled the population with descendants. Their sons brought wives, and new genes, from outside villages. When we arrived in 1999, one of the founders, "Pa" Michel, still lived in Bikol, but he spoke no French and had gotten quite old. Chief Gaspard, the prominent voice of the village when we first arrived in Bikol, was the son of the other founder. His father's tomb was prominently displayed in front of his house.

Although the people of Bikol did not possess a deed to the land, which officially belonged to the national government, no one disputed their "traditional rights" to use it. We would have to negotiate with them if we wanted to put a chimpanzee sanctuary here. We would need to convince them that the chimpanzee sanctuary would improve their lives, and we would need to make it true.

Kenneth drove Estelle and me from Yaoundé to Bikol for an official meeting with the village community. We pulled into Bikol with every available space in the Pajero crammed with bags and boxes of food and drinks. We brought a hundred pounds of rice, a Styrofoam ice box of mackerel, large bags of bitter greens and peanuts for a local dish called ndolé, ripe plantains, and basins of tomatoes, hot peppers, and other spices for making sauce. I had wanted the feast to be vegetarian, but Kenneth, and even Estelle, insisted it would insult the community and the local government officials who would attend. Kenneth argued that we should take live chickens so we wouldn't have to worry about the meat spoiling. Estelle thought it was a practical idea, which even I could concede, but she couldn't bring herself to argue for traveling with and getting to know chickens that we would kill in Bikol. I finally compromised on bringing the dead fish in a cooler. To drink we brought several cases of boxed red wine, crates of beer, and soft drinks. There was also candy for the children. We decided it was prudent to keep the drinks locked in the Pajero, but we handed over all the food to the women of Bikol so they could begin preparing the feast that would top off the next day's big meeting.

A loving moment with Launa, who had been kept on a chain until we rescued her.

◀ Dorothy, captive and miserable for decades at Luna Park.

▼ Dorothy, finally free of the heavy chain around her neck.

Nama, enjoying a cigarette, entertaining tourists.

Nama, in the fresh forest air, observing the world from a tree branch.

Nama, napping in the sunshine.

Jacky, trapped in a cage at the Atlantic Beach Hotel.

Jacky, once deemed "insane," found peace at the sanctuary.

The alpha couple, Nama and Jacky, in a warm embrace.

▲ Visiting Pepe despite a warning sign on the cage at the Atlantic Beach Hotel.

▼ Pepe had been easy to love from the start.

Becky, delighting in a stuffed panda that Sheri and Edmund had brought to her.

Pepe grooms Sheri through the bars of his cage at the Atlantic Beach Hotel.

Becky, the mischief maker, in a tree at the edge of the forest.

▲ The Atlantic Beach Hotel, where Jacky, Pepe, and Becky were kept in small cages for so many years.

▼ Relaxing near Jacky, Pepe, and Becky in their satellite cage, at Sanaga-Yong Center.

Aerial view of the camp from a helicopter, set in a remote part of Cameroon.

A raffia cover keeps Sheri's and Kenneth's sleeping tents cool and dry.

Kenneth Fonyoy, Sheri's driver and translator, with the ever-faithful Pajero.

Sheri with Edmund, who was pivotal in getting the sanctuary started.

Sheri *(center)*, French volunteer Laurence Vial *(left)*, and Estelle Raballand *(right)*, just after testing a chimpanzee for TB at the Atlantic Beach Hotel.

Sweet and smart Simossa, who was raised around humans, needed a gentle introduction to her chimpanzee community.

▶ A drink and a laugh with
  Chief Ibraham of Mbinang
  on New Year's Day.

▼ A girl from Bikol 2,
  bringing bananas for
  the chimpanzees.

Kenneth with Mado and Gabby. The chimps loved him.

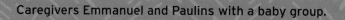

Caregivers Emmanuel and Paulins with a baby group.

Baby Bouboule, who was later adopted by Dorothy.

Bikol and Gabby, the first babies rescued, at the Sanaga-Yong Rescue Center.

Waiting for needed traveling papers just after officials confiscated Caroline from a hotel.

Gabby, playing like a typical chimpanzee baby.

Dorothy tenderly grooms Nama.

Chief Gaspard gave us permission to set up Estelle's small tent, which she and I would share, near one end of the village on a bumpy patch of dirt in a clearing between the last two stick-and-mud houses. For Kenneth's accommodation I had planned to rent a bamboo bed in one of the mud houses, like Estelle and I had done previously, but he preferred to sleep on the reclined front seat of the Pajero instead. The village of his youth was in a more affluent area of Cameroon's Northwest. The house where he grew up was made of mud bricks that had been formed in a brick press, not just mud plastered on a stick frame. The floor of his village home was made of concrete. For sleeping, Kenneth preferred the car seat to Bikol's dirt-floored huts.

At six o'clock on the morning of the meeting, when Estelle and I came out of our tent to meet the day's new light, we found the men of Bikol already congregated on their benches around a small fire, which was serving to chase away the slight chill of the early morning. They were joined by some of the younger children gnawing on cassava tubers that had been roasted on the fire. Meanwhile, the women were already at work. Using wood they had collected from the forest the previous day, they had already built two open fires in front of adjacent houses across the road from our tent. They were organizing various food items around the fires. They spoke to one another and gave commands to older, helping children in the Bamvéle dialect, or maybe it was Bobilis, which had been brought to the village by some of the wives. They sounded exactly the same to my ear. All the women worked, but no one rushed about. It was a calm but serious atmosphere in this important women's sphere of Bikol.

The women generally did a disproportionate share of the work in the villages. They cooked, washed clothes, cared for children, and farmed. The men hunted and worked beside their wives on the farms, but it was clear they had much more leisure time.

Estelle, Kenneth, and I left early for the town of Minta to pick up the divisional officer, Mr. Ndang Ndang Albert. His presiding over the meeting would legitimize it for the village people and make it legal for the government. Because we wanted to meet with the D.O. before the community meeting, Estelle and I both accompanied Kenneth on the four-hour drive to Minta. We wanted to make sure he understood that we were a very small nongovernmental organization (NGO), and our primary mission was conservation, not community development. I was very concerned about leaving people in the village with the impression that we would provide more for them than we could actually deliver. I would be the one answering to people in years to come.

Soon after Estelle and I had found the D.O. in Minta six weeks earlier, we had sent Kenneth by public transport to deliver a letter requesting his attendance at our community meeting. In his reply letter, which Kenneth had carried back to us, Mr. Ndang had selected this day to fit his schedule. When we arrived at his office, we found him waiting. He wore the short-sleeved, smartly pressed, khaki-colored uniform provided by the government to officials of his level. During a quick meeting in his office, Estelle reiterated the limitations of what we would be able to provide the community, and he assured us he understood the importance of not overstating our case during the community meeting. As we all headed for the Pajero where Kenneth waited, I had my first introduction to a Cameroon cultural requirement I hadn't considered before.

The D.O., like any self-respecting government official, would be obliged to go to Bikol with an entourage, the members of which were already waiting with Kenneth. As was the tradition, the uniformed captain of the military police, who was based in Minta, would go along to provide security. Mr. Ndang's assistant, in a starched white shirt and dress pants, would be necessary to record minutes of the meeting. Minta's "chief of post" for the Ministry of the Environment and Forestry (MINEF), a younger man in the hunter-green coveralls provided by the ministry, would come because the issue of a wildlife sanctuary was in his domain. The mayor of Minta, in suit and tie, would complete the D.O.'s entourage (for prestige, I supposed).

No one but Estelle and I seemed the least bit concerned that we would be packed like sardines in our smallish SUV for the long, bumpy ride back to Bikol, and other than glances of dread at each other, we didn't express our mutual annoyance over it. I directed the D.O. to the front seat, the place of honor, which left six of us in the back half of the car. Estelle opened the rear door to the small compartment behind the backseat, revealing our store of drinks still hoarded there. She shifted them around to maximize space for sitting and explained that someone would have to sit on a crate of beer. No one from the D.O.'s entourage spoke, but all heads turned toward the slim, young MINEF man. To show concern for his comfort and to emphasize Estelle's point, I slid out a flattened cardboard carton wedged against the side of the compartment, placed it on top of the beer crate, and smiled at him. He shrugged good-naturedly and crawled in. The remaining five of us tried to crawl onto the backseat, but it wasn't going to work. The D.O. called his assistant to squeeze into the front with him, which I worried would interfere with Kenneth's driving, but I couldn't see another alternative. Estelle and I hugged each of the back doors with the two men sandwiched between us. Ours was an unpleasant ride on the very bumpy road, wedged tightly against the unforgiving plastic of the doors, but we arrived back in Bikol just before three o'clock to greet a delighted population. A few people from the neighboring villages of Bikol 2 (Mr. Ngong Bipan Antoine was a relative of Chief Gaspard who had at one point joined him to settle in Bikol, but some disputes arose that caused the current chief, Chief Antoine, to found his own village of Bikol 2 a mile and a half down the road) and Meyene also came for the meeting. The attendees totaled about fifty, not including children.

For all the time and trouble it took to prepare, the community meeting was surprisingly short and simple. We met in Chief Gaspard's small house, where the five guests from Minta, Estelle, Kenneth, Chief Gaspard, Pa Michel, and I sat on bamboo benches arranged along three walls, with the D.O. sitting opposite the open wall. People squeezed into the room to stand facing the D.O., or they listened from just outside the house. Mr. Ndang spoke in French, and Pa Michel's son Samuel translated what he said into Bamvéle for people who didn't speak French—primarily older people—while Kenneth translated for me. During the meeting, it was decided that we would work with the Bikol community to delineate the sanctuary boundaries. Villagers would abandon any farms located inside the boundaries, and we would give individual farmers "symbolic" compensation, which meant that they would accept whatever we offered to relocate their farms. There would be no hunting with guns or traps inside the sanctuary boundaries. We, in turn, would provide jobs and buy fruits and vegetables from village farmers to feed the chimpanzees, thereby benefiting the community. Mr. Ndang explained that we were a small NGO, which could not solve all the problems of poverty in the community, but he added that our presence might bring tourists and other NGOs—which could, in time, bring more benefit. Several of the villagers gave speeches expressing the hope and belief that they would be better off with us in the community than they had been without us. Chief Gaspard thanked us for choosing Bikol over any other village. In conclusion, Samuel, the translator, thanked God for bringing us to them. And that was it!

After the meeting, the women delivered food on metal plates to the guests from Minta and the others of us seated inside. Kenneth acted as our drinks waiter—fulfilling the requests for wine, beer, or Coke from the special guests first, then staying outside to oversee the distribution of drinks to the population. The free flow of alcohol helped guarantee everyone's patience while they waited for food. It was Cameroon tradition that the important people should eat first, but I knew there would be enough food for everyone.

Around six P.M., Estelle and Kenneth left to return the D.O. and his entourage to Minta, where they would pass the night in the small hotel where Estelle and I had stopped for a bucket bath weeks earlier. The next day, Kenneth would drive Estelle back to Yaoundé. That night with the village population, I drank boxed wine and danced barefooted in the mud to an African beat pounded on cowhide drums by two young men from a neighboring village. Still unable to have even the simplest conversation with the people of Bikol, I nodded and smiled when anyone spoke to me. A burst of laughter often followed, and I wondered to what in the world I was agreeing. I didn't remember many names at that point, but people seemed genuinely happy to have me there. A group of several women joined forces to coach me on how to properly pulsate my hips to the drumbeat. My lighthearted but sincere efforts, emboldened by the wine, fell woefully short until I concluded that one had to be born to the special talent. However, in years to come, I *would* get a lot closer to this pulse of Africa and feel flattered when anyone said I danced like an African. When I realized I was quite drunk, I bade farewell with a wave to my strange fellow merrymakers, crossed the couple dozen yards to my tent, and zipped myself inside. The drumming soon stopped, and I drifted off to the sound of conversation and laughter.

Forty-eight hours after he left, Kenneth returned to Bikol. He and I camped in the village for the rest of June. Kenneth may have been a "city boy" as I teased him, but somewhere along the way he became committed to our cause, or to me, or to both—I was never sure what really inspired him most, but whatever it was, he worked beside me in the forest like the project was his.

Few drivers or translators in Cameroon, or anywhere else in the world, would have done the same. With Kenneth by my side, we explored many acres of forest behind Bikol. I tucked my pants in my socks, just above my ankle-high hiking boots, as some protection against ants, and Kenneth wore a pair of green rubber boots I had bought him in Yaoundé. There were several species of poisonous snakes in the forest, and I tried to watch where I put my feet so I didn't step on one. We saw the molted skin of a snake that I recognized from books as belonging to a deadly Gaboon viper. We both wore long sleeves as protection against mosquitoes and the ubiquitous small black biting flies, which swooped silently and relentlessly around our faces and sometimes got into our eyes. Once when I put my hand against a tree for balance, a piercing pain in my palm caused me to pull it back quickly. I never saw what stung me, but I watched where I put my hands after that. I had hiked a lot in the temperate rain forests of Oregon, but this African forest, while incredibly beautiful for its diversity of plants and animals, was a more hostile environment for humans. But we frequently saw small monkeys in the trees above us, and it punctuated the tedium of tiny annoyances with an exciting sense of exoticism. Through it all, I tried to envision the future chimpanzee sanctuary and plan its layout while Kenneth wielded a machete, cutting just enough to clear our access to various vantage points. Our goal was to choose our future campsite and the exact location of our first chimpanzee enclosure.

My résumé in no way qualified me for the job. My training was in veterinary medicine. The most complicated structure I had designed and built was a rabbit cage for my backyard in Beaverton, Oregon. Somewhere I had grabbed on to the audacious conviction that I could and would construct a chimpanzee sanctuary from the ground up in this remote part of one of Africa's most difficult countries. I hid a fear of failure that lurked just beneath my façade of self-assurance. From my fear, and from my constant awareness of the intolerable consequences of failure, grew a single-minded determination. The chimpanzees I had grown to love would not end their lives where they were. I would give them a different story if it killed me.

At the end of each day in the forest, Kenneth and I dragged our sagging, dirty selves back to Bikol. Madame Beatrice, one of the older women in the village, usually carried water from the river for me to use for my bucket bath and even heated it over her fire. We couldn't speak to each other, but her simple act of kindness to me in those early days spawned affection between us that lasted until she left the village years later to go back to the village of her childhood. Each afternoon, Kenneth disappeared for half an hour and came back looking clean. I wondered if he had a woman who was helping with his bath, but as far as I knew he always slept in the Pajero alone.

I hired Colbert, who had been Estelle's and my guide during our first trip into the forest, to make a bamboo bench. I placed it a yard away and perpendicular to the vertical zippered entrance to my tent, a psychological blockade of sorts. On the other side of the bench, I arranged rocks, recently unearthed when the bulldozer made the road through Bikol, in a circle to create a small hearth. Each night Kenneth and I quickly built our fire using borrowed wood embers from one of the other village fires. With three larger rocks, we made a triangular shelf over the fire on which to set our cooking pot. We cooked instant Ramen noodles or boiled yams with eggs. Dessert was an avocado or a banana. We were both too exhausted to be creative in the realm of campfire cooking. We usually ate silently, gazing at the crackling fire.

I was comfortable enough on the foam mattress in my tent, until one night about ten days after we arrived I woke to the sound of heavy rain and noticed water accumulating in the corners of my little domed room. As the downpour continued, water kept seeping in, occupying more and more space around the edges of my tent floor, pushing me to a smaller and smaller island in the center. I was desperate to the point of tears for sleep and could no longer lie down because my mattress was soaked on both ends. My raincoat was in the Pajero, but finally I made a run through the still pouring rain to cover the seven or eight yards from tent to car. I yanked at the front, passenger-side door handle, only to find that the door was locked, and had to pound and shout at the window through the noise of the rain to awaken a soundly sleeping Kenneth. Soaked to the skin, I entered the car and finally slept in wet clothes on the folded-down passenger seat beside him.

The next day, while I worked to dry out the tent—over and over soaking up water with a towel inside and wringing the water from it outside—Kenneth and some of the men from the village organized themselves to erect a roof of palm fronds over it. Others dug a trench around the tent, which would drain water away from me during future rains.

One of the village latrines was located about twenty yards behind my tent. It was a hole about seven feet square, crossed by a number of sticks on which a person could stand. It was among the trees at the edge of the forest, but there were no walls, or even dense foliage, around it. Early one morning I rushed out of my tent and headed to the latrine, preoccupied as I often was, and somehow neglected to actually look at the latrine until I was within ten feet of it. A young man from the village was squatting over it. His smile and casual wave were spontaneous and relaxed.

"Bonjour," he said.

"Bonjour," I parroted, as I jerked my head away. In my one-sided embarrassment, I made a hasty retreat to wait behind a tree until he had finished his business and gone. My concept of privacy and my need for it were out of place in the village.

When all the items in my forest wardrobe became as dirty as I could stand, I accompanied a young woman called Emilienne, who was the wife of Samuel and daughter-in-law of Pa Michel, and her eight-year-old son, Mesmin, down the steep slope of a ravine to the small Ndian River. The people of the village washed their clothes and bathed in the Ndian. They also collected water from it for drinking and cooking. Grasping the thin metal wire handle looped across the top of my black bucket, I carried down my dirty clothes and square of brown soap, just like Emilienne did. When I saw her dipping a dirty item of clothing directly into the stream, I did the same with a pair of pants. This is where our aptitude diverged sharply. With clumps of cloth in each fist, I rubbed sections together, trying to remove the considerable amount of visible dirt, spot treating the fabric with my block of soap as I deemed necessary. Aware that it was polluting the stream, I was trying to minimize my use of soap. Meanwhile, I saw that Emilienne was vigorously pummeling her clumped-up, soapy article of clothing on the smooth top of a large gray rock, which protruded from the shallow water near the edge of the stream. She had stripped down to a thin tank top, and though she was a tiny woman, I could see the well-developed muscles of her back and arms ripple and shimmer under her chocolate-brown skin. Her fitness came not from a gym, but from hard work, and her competence from years of uncomplaining repetition. I thought she was beautiful, and I couldn't help watching her. She rotated and pounded sections of cloth, splattering thick white lather all over the rock and into the stream.

When her glance met mine, she smiled and gestured to another big rock in the stream, near where I sat. Okay, I nodded; I would try it her way. My necessity for clean clothes was trumping my conscience on the use of soap. However, when my futile effort soon ended with two of my knuckles scraped and bleeding, I went back to *my* method, the way I had very successfully washed bras in the sink all my adult life without injury. When Emilienne had finished her load of clothes, twice as many as I had brought, she gathered my remaining two dirty items, and, seemingly without judgment, she washed them for me.

———

When all the clothes washing was accomplished, she and Mesmin stripped and she used her block of soap to thoroughly lather both of their bodies from head to toe. My bath in the cold stream was less thorough. When the three of us had dressed, Emilienne filled with water a big aluminum basin Mesmin had carried down from the village—first dipping it in the shallow stream to scoop up as much water as possible, then splashing in water with her cupped hand to fill it almost to the top. She squatted down, and Mesmin helped her heft the basin onto her head without spilling a drop. She struggled only slightly to straighten her knees and stand under the heavy weight of water. As she stood balancing the water-filled basin with one raised hand, Mesmin placed a bucket of clean clothes in her other hand. He filled one of the much smaller black buckets with water to place on his own head, then carefully bent to pick up their remaining bucket of clean clothes. They worked in perfect, practiced collaboration without speaking a word. With their loads of water and clothes equitably and practically distributed, they started up the steep incline. I followed with my bucket of relatively clean clothes, embarrassed that I hadn't brought an extra bucket to even try to carry water. Even without the weight of water, my legs ached from the exertion of the ascent.

After that first trip to the Ndian, I hired Emilienne to wash my clothes when they were dirty, an arrangement with which we were both delighted.

To Kenneth's relief, and mine, too, I finally selected a mostly abandoned cassava farm about a mile into the forest behind Bikol as our campsite. Madame ("Ma") Clementine, an elderly woman from Bikol, had managed to slash and burn away the trees from the site many years before, but she was no longer able to work the farm. It was mostly overgrown with weeds and small trees, but I would give her compensation for it anyway.

Next we had to make the path that would become our driveway. With our compass and machete, Kenneth and I cut a mostly straight track from a point on the road about a quarter mile from Bikol through the forest to our future campsite. As I referred to the compass and pointed the way with my outstretched arm, Kenneth walked a few paces ahead of me, cutting the track with his machete. I struggled to watch for ants and snakes and to keep my footing on the uneven forest floor, while keeping my arm pointed in the right direction for Kenneth, who looked back every few seconds to see where my arm was pointing. It was exhausting work for both of us, undoubtedly more so for him. I frustrated him by shifting the line several times to minimize the number of trees that would come down when the bulldozer blasted through.

When I was satisfied, we spent a whole day marking trees along the route with a big red X using paint I had brought from Yaoundé. Again and again, with a deep sense of the wrongness of it, I marked for destruction beautiful trees of many species—some huge and ancient, others young and perfect. In the cool, oxygen-rich forest, I could feel the life in all these trees—so benevolent and giving.

"In the long run our presence will be good for the forest and the animals who live here," I told Kenneth, justifying as best I could both to him and myself.

Kenneth began to tease me unsympathetically. "We may be 'In Defense of Animals,' but we're 'In Destruction of Trees,'" he said, making an unfunny joke.

I fought momentary compulsions to drive away from this place and leave these innocent trees as I was finding them. But of course, my empathy ran deeper for the chimpanzees, and I knew that even if we left, the forest would not remain exactly as we were finding it. Jean Liboz would be logging this area if we weren't here.

While we were camped in Bikol, Liboz and his crew were working five miles away building a bridge across the Yong River, close to where the train tracks crossed. His bulldozer had cleared a dirt track wide enough for a vehicle to pass from Bikol to the bridge, and another ten-mile track on the other side of the river to the town of Bélabo. On sunny days when it had not rained the night before, our Pajero could cross the unfinished bridge and pass to Bélabo for supplies. When it did rain, even four-wheel-drive vehicles bogged down in the mud on the fresh raw track.

The small town of Bélabo hosted the only market within reasonable walking distance where the people from villages on our side of the Mbargue Forest could sell the produce they grew. Many years earlier, the government had built a wooden bridge across the Yong and a road from it to Bélabo. However, termites had destroyed the bridge, and foliage and trees had recaptured the road long ago. When Liboz arrived on the scene, only remnants of the old bridge and a well-worn human footpath remained. The women and older children walked from the villages all the way to Bélabo carrying plantains or basins of cassava, yams, or peanuts on their heads, traversing the river on the train tracks. Not surprisingly, the village community was happy about Liboz's logging road and bridge, because it meant access to public transport—motorcycles and bush taxis—and maybe other kinds of development in the community.

One day around noon, Kenneth and I came into Bikol from the forest to have some lunch. He wandered away toward the opposite end of the village, probably seeking a few minutes of anyone's company but mine. I was alone on my bamboo perch in front of my tent when I saw Chief Antoine, the leader of the village of Bikol 2, to whom I had been introduced at our official meeting with the divisional officer, limping rapidly down the center of the road. During our first meeting, the chief had shown me what remained of his left foot; about a third of it had been destroyed by a Gaboon viper bite ten years earlier. He said it was the reason he had requested a job at the sanctuary for his son-in-law and not himself. When Chief Antoine spotted me on the bench, he veered rapidly toward me. It seemed he had come to find me. I stood and shook his hand, trying to smile in a welcoming way, until I realized that he was extremely upset. He spoke rapid French, unintelligible to me, gesturing frantically to his crotch as he spoke. When I bent over to get a closer look at the problem with his blue polyester-clad crotch, he stepped back abruptly, stopped gesturing to his crotch, and increased the volume and emotional intensity of his indecipherable plea. He was on the verge of tears, and now he was waving his hands in the direction of his village, the direction from which he had walked. At this point in the futile exchange, I turned my head toward my last sighting of Kenneth, circled my mouth with my hands, and shouted his name as loudly as I could, not knowing if he was within earshot. He was not, but I heard a man's voice several houses away echo "Kennet"—the village people never pronounced the *h* in his name—much louder than I was capable, and then another fainter voice shouted it again, farther away down the village communication grapevine. Within a few seconds, I was relieved to see Kenneth running our way, across the bare dirt that fronted the row of tiny houses. As he neared and saw that I was still standing and appeared well, his pace slowed and his face relaxed visibly.

"I need translation!" I sputtered at him before he had come to a complete stop. I briefly regretted alarming him, imagining him imagining me bitten by a snake or attacked by a wild animal. He told me later that, knowing I was clumsy, he only feared I had fallen in the latrine.

As Kenneth translated, Chief Antoine explained, obviously exerting great effort to stay calm, "My daughter walked to Bélabo to sell our coco yams six days ago. Five days ago on the way home, she gave birth to her baby daughter on the trail. My daughter was fine until this morning when she started bleeding." Again he made large flowing gestures with both his hands in the vicinity of his crotch to show *where* his daughter was bleeding *a lot*.

I absorbed the fact that this girl, nine months pregnant, had walked from Bikol 2 to Bélabo (seventeen miles) carrying a heavy basin of produce on her head and had given birth on the ground on her way back to the village.

"Is the baby okay?" I asked incredulously.

Chief Antoine shouted in teary exasperation at my obtuseness, "The baby is fine! It's my daughter who is dying!"

"Let's go," I said to Kenneth.

We bumped and slid through the mud from Bikol to Bikol 2, about a mile and a half, as fast as we could risk going. Fortunately, it hadn't rained in more than twenty-four hours, and the sunny morning had helped to dry the roadway, but without caution we could have easily slid into the mud bog along the sides. When we reached his village, Chief Antoine led us into a house, past an expectant throng of village women gathered near the portal. Along one wall of the ten-by-ten-foot room of the little mud house, I saw a young woman lying on a thin foam mattress atop a narrow cot of bamboo. An older woman, obviously her mother, who was sitting on a small bamboo stool beside the cot, moved out of my way as I approached. Kenneth stood with Chief Antoine against the opposite wall. I sat on the stool vacated by the mother and looked at the girl, who was barely conscious. She looked like a teenager, or at most in her early twenties.

"Fait ça (*Make this*)," I said in bad French, and opened my own mouth wide to model what I wanted her to do. She slowly opened her mouth as wide as she could manage. Her tongue was white. I nodded, and she closed her mouth. I lifted her upper lip with my finger to see her gums, hoping for a comforting hint of pink. Instead, her white gums rattled my nerves. The girl really was dying.

"What is her name?" I asked Kenneth, buying time while I considered what to do.

"Vivian," her mother answered.

I sighed and stood up slowly, willing myself to project a calm demeanor.

"We'll try to get her to the hospital in Bélabo," Kenneth began to translate, but they had already understood me say Bélabo.

"Merci Dieu! (*Thank God*)," the chief said, bowing his head in relief. Other than Liboz and his men, who were working five miles away, we had the only vehicle in the area.

There was a government hospital in Bélabo, and it sometimes had a doctor. The facility was poorly equipped and dirty, but it offered the best chance of survival for this young woman. A month earlier, before Liboz had opened the road to Bélabo and before *we* were there with a vehicle, Vivian would have died here in the village. Even now, I didn't think there was much hope for her. First, she would have to survive long enough for us to get her to Bélabo, and the chances of that would be much better if we could avoid getting stuck in the mud on the way. If we got as far as Bélabo, there were a number of grim factors pertaining to the quality of care that would be available. That I was pessimistic didn't matter. We had to try.

Kenneth carried Vivian and gently placed her on the backseat of the Pajero. Chief Antoine and his wife, whose name I learned later was Cressance, entered on either side of their daughter. Cressance carried a zippered plastic market bag, which she had evidently packed earlier in desperate hope that her husband would find transport to a doctor. Another middle-aged woman carrying an open-top plastic bag with plantains sticking out—they would need to eat at the hospital—entered the car beside Chief Antoine. I decided that Kenneth would drop me off in Bikol and proceed to Bélabo without me.

While Chief Antoine relocated to the front seat, I ran to the tent and brought 10,000 CFA, about $20, back to Kenneth in case he needed it. The subject of money had not come up with Chief Antoine, and I assumed, naively, that he had the money to pay for Vivian's care. Our contribution would be the transport. Standing out on the dirt beside the car, my eyes locked on those of Cressance, who turned her head to continue meeting my gaze as the Pajero pulled away. She was amazingly stoic, and other than to tell me her daughter's name, she had not spoken to me. It was through her eyes she showed me her pain. I was not religious, but as I watched the Pajero pass out of sight I said a silent prayer to God that this woman would not lose her child.

Kenneth returned to Bikol around eight o'clock that evening. They had found the doctor at the hospital, and Vivian was admitted for treatment. Chief Antoine ran out of money in their first hour at the hospital—all medications had to be purchased prior to treatment—leading Kenneth to donate my 10,000 CFA before he left. Kenneth had other bad news.

En route to Bélabo, he had seen Liboz. We had informed him a few days earlier that we had finished marking the path of our future driveway and were waiting in Bikol for the bulldozer. He had just informed Kenneth that the machine was broken, and it would take at least a week to get it repaired. This was distressing news for me. We needed to finish the first enclosure and bring Jacky, Pepe, and Becky before the heavy rains in September would make some of the roads impassable. I hoped to bring Dorothy and Nama soon afterward, although we still didn't know what the proprietors of the hotel would decide. In our part of Cameroon, the rainy season extended from April to mid-November, but July and August usually saw less rain. This was our opportunity to move forward with construction, and I constantly felt the pressure of the ticking clock.

We could accomplish no more in the forest until we could transport building materials to a cleared campsite, so I decided that we would return to Yaoundé for a few days. Kenneth and I were both tired, and I dreamed of a warm shower. More important, I wanted to check on Dorothy and Nama, which I could do on the way to Yaoundé. With nothing urgent to accomplish in the forest, we were eager to go. We packed out at dawn, leaving the tent in place for when we returned.

Unfortunately, it had rained the night before. The truck slid precariously along the mud-slick road, the accumulated slime on the tires effectively obliterating the usefulness of the tread. I welcomed the frequent water puddles because they washed mud from the tires, but unfortunately gave us only short-lived advantage. About two miles from Bikol, Kenneth lost control going down a hill and the truck slid into the mud bog along the edge of the road. We had a shovel and a machete, with which the two of us worked for over an hour trying unsuccessfully to dig ourselves out.

Finally, a young man pushing a bicycle through the mud met us coming from the opposite direction. He introduced himself as Assou Francois, Vivian's husband and Chief Antoine's son-in-law. We were delighted to learn that Vivian was still alive in the hospital and doing better. Assou wholeheartedly joined our effort to liberate the Pajero, and soon two other men from the nearby village of Mbinang pitched in too. With another hour of digging and pushing, the five of us were able to free the car. During the meeting with the D.O. several weeks earlier, Chief Antoine had requested a job at the sanctuary for his son-in-law. This happy young father with the surviving wife must be the one of whom he had spoken. I hoped this spirit of collaboration between us was a sign of things to come.

eight

NOTHING WORKS,
BUT IT ALL WORKS OUT

Rolling down the Luna Park Hotel driveway always brought on the same nauseating anxiety. Would Dorothy and Nama both be alive, no worse than I had left them? I sighed in relief to see Dorothy, and then a few moments later Nama, who was standing on two feet, using a rake to clean her dirt. She had apparently taken the rake from the groundskeeper. They recognized me passing in the Pajero, and both watched eagerly as we parked and got out.

I had returned several times since our first visit, sometimes with Estelle, sometimes, like today, with Kenneth only. It had been almost a month since I had seen them last on our way to Bikol.

As I crossed the fifty-yard stretch of grass toward the tree where Dorothy was chained, she met me with her now familiar hand-flopping, screeching hysteria. I had known that it was an expression of desperate need mingled with hope, but it was Estelle who had known intuitively what Dorothy needed most. On our second visit, Estelle had taken a leap of faith and slid in between Dorothy's flopping hands to hug her. So on this day there was no mystery in it. More than anything, Dorothy craved contact; she wanted to be hugged. I squatted a couple of feet in front of her, and she quieted to welcome me as I duckwalked into her arms. I wrapped my arms around her and breathed in her distinctive musky body odor. As I patted her back soothingly, I felt her relax. After a few seconds, Dorothy pushed me out far enough to study my face and moved her mouth as though she would groom it. I felt her warm breath on my face, and happily anticipated her gentle grooming fingers, but she changed her mind and hugged me tight against her again. As much as I loved this intimacy with Dorothy, as much as I knew she needed it, I was aware of Nama watching and feeling left out. After another minute, I backed up to retrieve and hand to Dorothy the bag of bananas, papaya, mangoes, peanuts, and yogurt that I had packed at the market just for her. By now I knew she preferred to keep the bag and take her time exploring the contents.

I took an almost identical bag to Nama, who stood to take it eagerly and then sat down. After our first visit, I had diagnosed a heavy load of intestinal parasites and had given her two treatments already. She looked slightly more robust than when I had last seen her. Unfortunately, I couldn't hope to rid her of the parasites as long as she lived on that contaminated patch of filthy dirt. Nama explored the contents of the bag for a few seconds and fished out the small papaya. As she savored her first bite, she looked me in the eyes and put per fingers to the chain around her neck—as though to remind me. "I could never forget it, Nama," I assured her softly.

Kenneth and I were exhausted after our long trip from Bikol, and we hoped to get to Yaoundé before dark. After I fed the monkeys and gave water to everyone, we left, knowing I would visit again soon.

Dana and Estelle were on vacation in the United States, and I stayed in their detached guest room with access to their house and kitchen. While we waited in Yaoundé, I visited Dorothy and Nama twice more and rested in comfort while I worried about the delay we were facing.

Unfortunately, the same day we finally received a message from Liboz that the bulldozer was set to go, I got sick. A high fever, headache, and excruciating pain in my hands, feet, and knees rendered me useless for almost ten days. I sent Kenneth to guide Liboz's bulldozer through the forest to our campsite and worried constantly about the time passing while I lay there doing nothing constructive for days on end, nothing but thinking and planning.

As soon as we could transport our building materials into our campsite, we would build a satellite cage for Jacky, Pepe, and Becky and soon afterward a second one for Dorothy and Nama. They would sit adjacent to a tract of forest we would eventually enclose with solar-powered electric fencing. I would take the chimpanzees back to the forest, not to live in cages, but to feel the earth beneath their feet and to view the horizon from the forest canopy. Electric fencing was the only way to give them their piece of natural habitat. I had seen electric fencing used for monkeys in Texas years earlier at Limbe Wildlife Center. I had studied the system at the Pandrillus sanctuary in Nigeria, which Edmund and I had visited on our way to Cameroon. Peter Jenkins, codirector of Pandrillus and Limbe Wildlife Center, had taken time to explain how their complicated (or so it seemed at the time) electric fencing system worked.

I would be using the same equipment, which Peter Jenkins himself had transported to Limbe, via Nigeria, from the United States. Peter, who was a charismatic personality of many and varied talents, had managed to get a big load of equipment transported for free from the United States to Nigeria, through Mexico, on a charter plane. We contributed to the costs of shipping to Mexico, which were minimal compared to the price of commercial transportation to Cameroon, and Peter included our equipment in the shipment to Nigeria. Afterward, he brought our equipment, along with supplies he had bought for Limbe Wildlife Center, by boat from Calabar, Nigeria, across the Bight of Benin to Cameroon. He got the equipment as far as Limbe, and the rest was up to me. A Limbe businessman, a friend of George Muna, had allowed us to use one of his trucks to move it to the big fenced yard of the Coron sawmill in Yaoundé, where it all had been stored for two months. The equipment would take its final journey to the Mbargue Forest on one of Coron's logging trucks as soon as I was well enough to move with it.

Finally, after ten days that seemed like a hundred, I was well enough to return to the forest, although I would suffer from arthritis in my hands and feet for more than a year. I know now that the chikungunya virus, which is transmitted by daytime mosquitoes and is endemic in Cameroon and many other African countries, probably caused my symptoms. African forest primates are reservoirs for the virus, as are humans. It's likely that I was infected with the virus by a mosquito bite while I was trekking through the forest.

Gingerly, on sore feet I walked through the sawmill yard monitoring the loading of our imported fence equipment and lots of other building materials we had bought in Yaoundé. We had hired a welder to create large latticework panels from individual iron rods and then had stored the panels at the sawmill. Welded onto a sturdy frame, these would form our cage walls. A company in the Douala port had donated a twenty-two-foot metal shipping container, and we had managed to transport it, too, to the Coron sawmill weeks earlier. When all of it had been loaded and secured on the long flatbed of the logging truck, we set off for the Mbargue Forest. A shipment of gold couldn't have been more valuable to me than those precious building materials. Because I wanted to make sure everything arrived safely, Kenneth and I left Yaoundé in the Pajero, following behind the big logging truck. However, an hour outside of Yaoundé, when the pavement ended at the town of Ayos, the dust stirred up by the wheels of the big truck on the dirt road, even in this season of not-infrequent rains, was blinding. We managed to pass the truck with the intention of driving in front of it, but this arrangement didn't work out well either. The washboard bumps created by heavy logging trucks limited the speed of our small Pajero much more than it did the heavy truck. The truck driver was impatient with the slow speed and made me nervous by driving too close to us. In the end, we let him pass and fell far behind him. He knew where he was going because he was waiting for us at the campsite Liboz had cleared when we arrived many hours later.

We arranged for truckloads of sand and gravel to be brought from Bélabo so that we could make cement by hand, using water we would carry from the river. Our first cemented structure was the floor of our pit latrine. Although its walls were made only of palm fronds, the wonderful privacy it afforded felt like a five-star luxury to me. It and a raffia roof over our side-by-side tents were the only structures we built before starting the cage. I had borrowed a second tent from Estelle and Dana for Kenneth. This would be the extent of our camp infrastructure, our base of operations, until after we brought the first chimpanzees to the sanctuary.

Because building the electric fence would take time—just clearing the fence line of trees could take months—I would focus first on building a big satellite cage. I thought we could build it in a few weeks, and it would be a big improvement over where the chimpanzees were living now. Then, as soon as possible, we would build the electric enclosure and get them back into the forest.

We hired our construction team of local village people. Mr. Francis, a middle-aged man from the village of Mbargue, on the far side of the Mbargue Forest, was our construction technician. He knew nothing about building cages, but he could lay cement and had built houses with wood. His was the highest level of skill we could find. Six men and teenagers from Bikol 1 and Bikol 2 were the unskilled laborers.

I lacked a blueprint for the cage. I had made rough sketches, I went over every detail of the three-chambered design in my head, and I perfected it as well as I could with my level of skill and knowledge. With all the human and sliding chimpanzee doors that had to be anchored in concrete, it was complicated construction, some of which would have been difficult for me to explain to Mr. Francis even if we had spoken the same language. Under the circumstances, the task was ridiculously difficult. I hovered over the men every minute of every day as they planted poles in concrete to make sure my amateur design was followed precisely. When Kenneth wasn't with me to translate, I drew in the dirt or pantomimed to clarify my intent.

None of the men on our team, with the exception of Mr. Francis, had ever held a paying job before, and having an American woman boss undoubtedly veered into the realm of the bizarre for them. Up to this point, I had enjoyed a sweet relationship with the people of the villages. Now, trying to finish this first satellite cage before the heavy rains made the roads impassable, I pushed hard for the work to move fast. Our cultures and work ethics collided head-on. I wanted to start work at seven sharp every morning to accomplish as much as possible before the day got hot, but it was rare for all the employees to arrive on time. I tried explaining the importance of arriving for work on time, but they were farmers and hunters for whom a schedule imposed by anything other than the growing season and the physical necessity of eating was difficult to adhere to. I got angry, but I had little leverage. Because I needed the whole crew to get work done, I couldn't afford to send anyone home without pay. When nothing else worked, I motivated them with bonuses for arriving on time.

For them to bring lunch with them in the morning was a cultural and logistical impossibility. Because women from the village cooked food for them and brought it whenever it was ready, usually late in the afternoon, I wasn't able to set a specific lunchtime near midday, which would have allowed us to rest and rejuvenate in the hottest part of the day. I always held out and ate whenever they were eating, and I noticed that our productivity decreased as the hours wore on without food or rest. I could have solved the problem by hiring a cook and providing lunch in camp. I was pinching every franc, but well-timed rest and nutrition might have increased productivity enough to make up for the cost of lunch. In any case, I didn't think of this obvious solution until it was too late. My narrow American perspective sometimes blinded me to culturally appropriate solutions.

One day around one o'clock, I went to the latrine in camp and got distracted by a discussion with Kenneth over the placement of a table in the little cooking area near our tents. When I came back to the cage site, all seven members of the workforce, including Mr. Francis, were sleeping under a big mango tree. To wake them, I used a near-empty plastic water container as a drum, pounding it furiously with a two-foot piece of cage metal. They all rose silently, rubbing their eyes and yawning, as they moseyed back to work.

Their typical slow walking pace, which was undoubtedly a sensible and lifelong adaptation to the tropical climate, contrasted sharply with my march to the drum of urgent purpose. That they could not seem to comprehend or respect my sense of urgency exhausted what little patience I had. When my frustration and stress compelled me to shout unintelligible words—not one among the workforce understood English—the expended energy was not entirely wasted because they usually understood that I wanted them to go faster. For a few minutes they would try to accommodate, or humor, me by working a little faster, all the while speaking in their language and laughing among themselves (at me, I was sure).

In spite of all the problems, work moved forward. I was watching the calendar constantly, trying to beat the heavy rains, while we worked around lighter, shorter rains that were already coming several days per week. Then came the problem of the generator.

Weeks earlier, in Bélabo, we had arranged to rent a big generator that could power welding equipment. Unfortunately, a bad surprise awaited us when we were ready to weld the standing poles and walls together. When Kenneth went to pick up the machine, the proprietor told him that it was broken, sitting idle, without an essential replacement part that could only be bought in Europe. Our search for another generator was a frustrating drama that led us to Bertoua, the closest big town, and then all the way to Yaoundé. I was almost out of money, which made the option of renting an expensive generator in Yaoundé less tenable. There had been so many costly contingencies for which I had lacked the experience to anticipate or plan. Edmund was doing his best to raise money in the United States, but with no accomplishments to show potential donors, it wasn't easy. I would learn later that he took out a loan for $2,500 in order to send me money when funds ran out.

One Saturday afternoon, I was alone at the cage while the work crew took their day of rest. Their Seventh-day Adventist religion prohibited work on Saturday. I always thought the village people embraced what they liked of the religion and left the rest, but try as I might, I couldn't convince them that God would smile on those who worked for chimpanzees, no matter upon what day of the week the work occurred.

"Madame, we can't put you before our God," Mr. Francis told me.

"Does he pay your salaries? Does he feed your kids?" I asked.

"He makes the food grow," he answered sincerely, shaking his head apologetically. I couldn't win this one.

That particular Saturday, fretting about the generator problem, trying to conjure up a solution, I was near desperation. As I sat cross-legged on the dirt by the cage, my troubled thoughts were interrupted by a lanky middle-aged white man peeking over my shoulder. Other than Jean Liboz, who I knew had recently departed for France, and a Catholic priest in Bélabo, I didn't know of any other expatriates working in the vicinity. I stood as Roger Odier, a middle-aged man with thinning light brown hair, looked at me through gold wire-rimmed glasses and introduced himself in English. He was a French national who had recently arrived in Bélabo to manage a wood transportation company.

"One of my workers told me about a crazy American woman working alone in the forest," he told me with a heavy accent. "I had to come see this thing for myself. You must have dinner with me at my house tonight and explain what it is you are doing here."

He got no argument from me. That night over a lovely dinner of green beans, white rice, French cheese, and red wine, I explained why I was building a sanctuary for chimpanzees, and I went into some detail about the threats they and other primates were facing in the wild. I made an effort to enunciate well and speak slowly as he got accustomed to my American accent. To his sympathetic ear, I confided my problems—most notably, I spoke with all the gravity in my heart about my urgent need of a generator to weld the cage together before the heavy rains would make the work difficult, and moving the chimpanzees impossible.

The very next day Roger delivered his company's generator, two huge electric spotlights that plugged into the generator, and one of his company's welding technicians to work alongside another welder I hired from Bélabo. Working all day, every day, and late into the night, we finished the cage in two weeks. I joked that

Roger had come to my rescue like a knight in shining armor, and I could tell he liked my corniness. I benefited from his platonic friendship for two years before he finally left Cameroon to join his wife and family back in France. During our final meeting, he told me why he had helped me. "You were hobbling around like a ninety-year-old woman with your sore feet, and you had more determination than I had ever seen. You touched something deep inside me." I might not have managed to finish that cage without Roger, and I said many silent prayers of thanks for the amazing good luck, or whatever it was, that brought this man and his goodwill to me at that pivotal time.

During the last week of August, the heavy rains came early, and, as a result, many of the roads were closed to big trucks. At the entrances to some of the roadways, policemen lifted fragile wooden barricades for smaller vehicles to pass while they turned back big trucks. At others, policemen weren't necessary, because cemented metal barricades were locked into place, leaving space for nothing bigger than cars and small pickup trucks to pass.

En route to Limbe, Kenneth and I stopped in Yaoundé for one night to rest and pick up Estelle. She and Dana had just returned from their vacation in the United States and France. Dana's hospitality and kindness to me in those days was beyond measure. He opened his doors to me whether or not Estelle was in the country, and I was a frequent dinner guest at his table.

Over dinner that night, Estelle surprised me by presenting a rational argument for waiting until the next dry season to move the chimpanzees. "You built everything in such a rush. It's better to make sure the cage is solid and all the infrastructure is in place before taking chimps there," she said, making a good point. "The roads are terrible now. What if we get stuck? It's not worth the risk to move them now."

I knew our infrastructure was minimal, and, having just passed over the muddy roads connecting the Mbargue Forest to Yaoundé,

I could understand why Estelle was worried. I too was afraid of it all going wrong, so I sat considering whether moving the chimpanzees at that particular time was worth the risk.

Sensing a lapse in my resolve, Estelle added, "What difference will three months make?"

Would three months make a difference? I wondered. Flashing through my head were the chimpanzees in Limbe—especially Jacky going crazier by the day—and Dorothy and Nama waiting on their chains. I concluded that three months would, indeed, make a difference for them. I had struggled for months against myriad obstacles, driven by one idea—I would move the chimpanzees as soon as possible. At this late stage, Estelle's coolheaded risk assessment, as well intentioned and logical as it was, could not deter me.

"I'll move Jacky, Pepe, and Becky within three days," I said, giving my decision. I was willing to move them with or without Estelle, but she never knew how relieved I was that she later agreed to come with me.

Once in Limbe, Estelle, Kenneth, and I faced the necessity of finding a vehicle that was both big enough to accommodate our three transport cages and small enough to fit through the road barricades. I had made the cages with a welder in Yaoundé and arranged their transport to Limbe weeks earlier. I had intended to rent a big truck into which the cages would fit easily, but it was too late in the season for that now. A truck would be too big to squeeze through the barricades and would be more likely to get stuck in the mud. George Muna had the great idea of renting a bush taxi, a type of van used extensively for public transport in Cameroon, and taking out all the seats to make room for the transport cages. We knew bush taxis could get through the barricades because they moved all over the country year-round. George donated the cost of renting the bush taxi.

To avoid the traffic jams, the throngs of people who would crowd around the van in excitement to see the chimpanzees, and the daytime heat of Douala and Yaoundé—the two big cities through which we would necessarily pass—we decided to begin the journey at night. Without traffic or other complications, we anticipated a fourteen-hour trip.

Shooting darts at chimpanzees is a horrible process that terrifies and infuriates them. I had darted Jacky, Pepe, and Becky before for tuberculosis tests, and Pepe and Becky had been open to reconciliation afterward. We had known each other longer and were better friends now, which for me made it seem a deeper betrayal. I hated it, but in 1999 we had no better options for getting them into the transport cages than to anesthetize them with blow darts. The night we were to travel, two experienced caregivers from Limbe Wildlife Center (LWC) joined us to help with the darting and with loading the chimpanzees once they were anesthetized. The yard-long pipe of my darting equipment could be connected to a compressed air pistol or used as a blowpipe. The LWC caregivers also brought a blowpipe. I was much better at using the pistol than I was at blowing darts, but the pistol could be more painful. Since Jacky was in the middle, I darted him first, so we wouldn't need to worry about him grabbing us as we darted the other two. It was relatively easy to use the blowpipe with Jacky, because he had been sleeping and I caught him by surprise. But from that moment on, it was a stressful affair. When the LWC team entered Jacky's cage to carry him out to the transport cage, Pepe and Becky barked and screamed aggressively at them. At first I think they assumed I was innocent, but once Pepe saw that I intended to dart him, he started barking at me and jumping around all over the cage to avoid the dart. I had to use the pistol to hit him because he was moving so fast. The whole aggressive interaction was incongruous with our gentle relationship. I finally hit him with the dart, though, and we got him into his transport cage. It was no easier with Becky. Left with no vocal support from Pepe or Jacky because they were both unconscious, she cried like a baby chimpanzee and held her arm out toward me for support, pleading with me to stop being mean. In the end, one of the LWC caregivers darted her with the blowpipe. I just hoped the chimps would forgive me.

Once all three of the sleeping chimpanzees were in the transport cages, we managed to line the cages up in single file from just behind the front seat all the way to the back door of the bush taxi. Estelle and I entered through the sliding side door and sat just beside the cages on the hard metal floor, while Kenneth took the front seat next to the driver, who came with the bush taxi. Even before we pulled away from the hotel, the chimps began to wake up from the light anesthesia. They were confused and terrified, but no one seemed to be holding a grudge against me specifically. They sought our comfort, which we offered as best we could in the cramped space that was uncomfortable for us all. At least they could see that we were all in it together, so to speak. This was closer than we had ever been to Jacky—the openings between the welded bars of the transport cage were small enough that we needn't fear him grabbing or biting us—and even he took some solace in our presence.

In Limbe, we had acquired documents from the local government authorizing us to transport the chimpanzees, but the size and temperament of our passengers dissuaded careful scrutiny of the papers at any of the many police checkpoints through which we passed. Although we were sometimes able to soothe Jacky and Pepe, and to a lesser extent Becky, there were many moments during our journey when none of them wanted to be consoled. Even within the small confines of their transport cages, they were able to mount frightening displays and were inspired to do so whenever strange people peered in at them from outside. To prevent the pinching of their feet or hands between the bottom of the cages and the ground or floor underneath, I had designed each of the cages on four corner feet. Too late, I was having second thoughts about the design. All three of the chimps quickly realized that rocking the metal feet of their cages on the metal van floor produced a satisfying accompaniment to their already deafening barking and screaming. No human voice, police or otherwise, could be heard above the intimidating racket. We were usually waved through the checkpoints quickly. I myself worried about whether the cages were strong enough to withstand the phenomenal strength of the chimpanzees. During every bout of frenzied cage shaking, I imagined the thousands of weld points that held the cages together giving way. As agitated as the chimpanzees were, I wouldn't be happy to have any of them loose in the van with us. I kept a syringe of anesthesia and the dart gun nearby, just in case.

The first six and a half hours of the journey were over paved road, and, at times, between police checkpoints, I fell asleep, but once we hit the bumpy, muddy road, which made up two hundred miles of the route, the discomfort for the chimpanzees and for Estelle and me increased dramatically. The cages bounced hard on the metal floor, increasing the anxiety of the chimps, who did not suffer silently, and causing me to worry that the feet of the cages, already making dents in the floor of the van, might actually break holes in it. Estelle and I bounced about painfully on the metal floor and both yelled for the driver to slow down, to little effect. From the floor of the van, we couldn't see the road to anticipate bumps. After an especially painful landing on my tailbone, I threatened Kenneth. "Make him slow down, or you're trading places with me!" Kenneth was persuasive enough, but soon speed wasn't the problem anyway. The muddy conditions of the road slowed us to a crawl, and still the van was slipping and sliding.

Then, it happened. About ten hours into our journey, the driver lost control as the van slid and came to an abrupt stop in the mud bog that ran along the edge of the road. His efforts to drive out only sunk us deeper in the mud. Following Kenneth and the driver, Estelle and I crawled out of the sliding door, sinking in mud to our ankles. The front and rear wheels on the passenger side were half buried in mud. I pointedly avoided eye contact with Estelle, but I could still read her thoughts, flying at me like barbs. "We'll get out," I reassured all.

Kenneth and I had had recent experience digging ourselves out of a mud bog. I knew it might take a long time, and it did. Estelle and I tried to comfort the increasingly distressed chimpanzees while Kenneth and the driver, both already exhausted from the night without sleep, took turns with our shovel, digging out mud from around the passenger side tires and creating a mudless trench of firm ground in front of each one. Finally, the driver decided to make a break. When he pressed lightly on the accelerator, the wheels found traction and rolled forward for two rotations only to bog down again. After all four of us pitched in to dig for another half hour, Kenneth, Estelle, and I pushed the van with all our might while the driver accelerated. The chimps contributed screaming and thunderous cage shaking to the tense drama. When the van finally jolted forward onto firmer ground, it sent me sprawling on my hands and knees in the red-brown mud, but as exhausted and relieved as I was, I hardly noticed. The three of us climbed in the van silently. We knew a celebration might be premature, since we still had a long trip ahead of us. Fortunately, although it was slow going, we weren't stopped again.

When we arrived at the sanctuary an hour before dusk, eighteen hours after leaving Limbe with the chimpanzees, the construction crew was waiting for us. The villagers had never seen live chimpanzees up close, and they were both excited and frightened. The satellite cage we had built for the chimps was a hundred yards from the camp down a forest trail . The crew helped us carry the transport cages, one by one, to the cage, and the chimps were mostly quiet for once. For the first time in their adult lives they could see and hear the forest around them, and though they had no idea what to expect and were undoubtedly frightened, they seemed fascinated at the same time.

The satellite cage had three chambers with sliding doors between them that could be left open or closed. Estelle and I briefly discussed whether to separate the chimpanzees in different chambers until the following day and decided against it. They had been through hell and the new environment would be foreign and scary. Hoping and believing they would seek comfort from one another, we left open the sliding doors between chambers and transferred the chimps into the cage. We didn't need anesthesia for these transfers. We used chains and padlocks to fasten the sliding door sides of each transport cage over a sliding door of the much larger satellite cage. When the adjacent sliding doors were opened, the chimpanzees could find relief in moving from the small, cramped cage to the much larger one.

We transferred Pepe first, and then Becky. The two long-separated "siblings" ran into each other's arms, screaming and grimacing—happy to be together, but unsure how to feel about the circumstances that were uniting them. I thought it was a picture of overwhelming emotional ambivalence. A few minutes later came the part we were worried about—how would Pepe and Jacky react to each other? In each other's arms, Pepe and Becky watched anxiously as we opened the doors for Jacky. When he ran toward them screaming, Pepe, the biggest of the three chimpanzees, stood upright to meet and support him in a mutual embrace. Becky and Jacky hugged too, amid more high-pitched screaming from everyone, before they all quieted and sat close, each comforted by the proximity of the others as they looked at their new surroundings.

Several minutes passed before anyone broke the huddle to explore the cage—a huge space by comparison to what they had known—and even then, they moved mostly together. Becky was first to venture through the sliding door and into the next chamber, only to rush back to beckon Jacky and Pepe to follow, which they did. After the three chimpanzees had moved en masse across the woodchip-covered floors of all three chambers, they started breaking away from one another to explore the more than four yards of vertical space with the sleeping and sitting platforms at various levels. After all the other humans had gradually wandered back to camp, I was sitting beside the cage on a patch of packed dirt in satisfied exhaustion when I heard the free-living chimpanzees vocalize from a distance. It was the first time I had heard them here. They must have been responding to the noisy commotion of these new chimpanzees in the territory. Jacky, Pepe, and Becky heard them, too, and rushing back together, they gazed anxiously into the quickly darkening forest, listening for the sound to come again. I had known the chimpanzees were living here in the forest—at least the villagers had assured me they were—but hearing them erased any doubt and made them real. Hearing the call of the free-living chimpanzees somehow made the lives of the captives seem all the more diminished. The best that I could give them wouldn't be good enough. I so wished that they could live truly free, that they were still living free.

The sad truth is that after a chimpanzee is orphaned and raised by humans, there's not an easy path back to freedom. On the one hand, released, or reintroduced, former captive chimpanzees face dangers in the forest that make survival difficult: fiercely territorial free-living chimpanzees who are particularly dangerous to strange males, and poachers with shotguns who can easily target chimpanzees who lack fear of humans. On the other hand, the same familiarity and lack of fear that make them more vulnerable to poachers can also make them dangerous to humans. Humans have impacted the lives of bushmeat orphans to a huge extent, and our involvement has changed our status. We are not merely a potentially dangerous and socially insignificant species, as we are to free-living chimpanzees. Their responses to us depend in large part on their individual exposures to human kindness or cruelty or indifference, but regardless of the emotional contexts, humans are part of the social experience of captive chimpanzees. Virtually all chimpanzees who grow up with humans know they are physically stronger than we are, and their sense of territoriality can extend to us.

In any case, older damaged chimpanzees who had relied on humans as long as Jacky, Pepe, and Becky wouldn't be good candidates for release. They would always live at this sanctuary, but very soon they would have their own small piece of habitat here.

The free-living chimps didn't speak to us again that evening, but, as darkness came, so did the nightly symphony of insects and other nocturnal wildlife, only a few of which I could recognize by their calls at that time. After their brief excursions up to the higher platforms, Jacky, Pepe, and Becky had settled down to sleep in a heap on the floor. When they were no more than dark shadows, slightly discernible only in movement, I hobbled along the trail by the sparse light of my dim flashlight back to camp.

We had taken the first irrevocable step. Jacky, Pepe, and Becky were here—a huge step closer to being back in the forest where they belonged. Their lives, and mine, had changed forever. There could be no turning back now.

nine

FORCED SEIZURE

In late October 1999, we finally heard from the manager of Luna Park Hotel that his family would allow us to bring Dorothy and Nama to our new sanctuary, but it would be another five months before we were ready to bring them. I was working as fast as I could to build a second satellite cage for them, but there were stressful complications and delays. When a cement shortage in the country—more accurately, a total lack of cement anywhere in Cameroon—stopped our construction for two months, I took a quick trip home to Oregon. I needed to work with Edmund to raise more money and look for someone with more expertise than I had to help me build the electric fence. Since I lacked electrical experience and a deep understanding of the way electricity worked, building the fence alone on the foundation of a crash course in Nigeria more than a year earlier had begun to appear daunting. I left Kenneth in charge of caring for the three chimpanzees with the support of two volunteers and our local staff of seven men, who worked as caregivers, groundskeepers, and night guards.

When we had asked for job applications for the six permanent positions at the sanctuary, seven men applied. None were more or less qualified than any other. They had lived in either Bikol 1 or 2 all their lives, had minimal education, and had never held jobs before. We couldn't arbitrarily turn one man away, so I hired seven men for six positions and created a schedule that worked. No women applied initially. Estelle suggested that it wouldn't have been culturally appropriate for them to compete with the men for jobs. I believed this was part of it, but I also believed they were just too busy with children and farming. Eventually we *would* have female employees, who were sadly childless or who had mothers or sisters to help them with chores in the village. Estelle had stayed in camp for a few days just after we brought Jacky, Pepe, and Becky and helped me train that first round of employees. The cook learned his schedule for preparing yams and sweet potatoes for the chimps and local food for the staff lunches (I had finally figured out that it would be easiest to prepare lunch for village employees), the gardener learned where and what to plant, the night guards learned that keeping ants out of the cage was their most important task, and the two caregivers, Assou and Akono, learned to feed the chimpanzees and clean the cages. I had built one section of one cage wall with smaller holes in the mesh, so we could hand-feed Jacky without risk. We had hand-fed him in the transport cage on our way to the sanctuary, and on his first morning at the sanctuary, when Estelle handed him his bananas, he took them from her gently. At the beginning, the caregivers learned to hand Jacky his food from his special section, but soon they were feeding him from anywhere in the cage just as they did the other two. During the training, Estelle didn't tolerate any tentativeness in the caregivers' approach to the chimpanzees, and she expected nothing less than vigor in their approach to cleaning chores. I thought they did their best to comply, and once they got over their initial fright, they seemed to be tickled by the novelty of being so close to big chimpanzees. When I heard them laughingly refer to Estelle behind her back as Margaret Thatcher, I couldn't help but wonder what they might be calling me, but I knew they harbored no malice. By the time I left for the United States, they all had some job experience under their belts.

While I was in Oregon my friend Matt Rossell introduced me to his brother Greg, who with his wife, Anita Phillips, operated a small construction business building and selling houses. Together, they could manage all aspects of constructing a house, and they understood electrical circuitry. For several years, they had been incorporating into their work schedules enough downtime to volunteer their construction expertise to projects they thought were worthy, and they both liked a challenge. To my great relief, Greg and Anita agreed to come to Cameroon for eight weeks and take charge of building our electric fence. They studied the electric fence built by Peter Jenkins for juvenile chimpanzees at Limbe Wildlife Center and then adapted the model for our adult chimps. Working at a phenomenal pace, they finished our first solar-powered fence within six weeks. Encompassing five and a half acres of lush forest, at the time the enclosure seemed huge to me.

Greg also helped finish the new satellite cage for Dorothy and Nama, so that by mid-March 2000, we were ready to bring them to the sanctuary. I rode the train to meet Estelle in Yaoundé, leaving Kenneth to manage the sanctuary. He was to follow me in the truck a couple of days later to pick up Dorothy and Nama. Estelle and I arrived at the Luna Park Hotel around midday, and Dorothy and Nama recognized us in her white Toyota Starlet as we drove past. We delivered their food and spent a few minutes visiting. I was so excited with the expectation that we would be taking them with us in a couple of days, and I so wished they could understand that their lives were about to change dramatically. They surely noticed that my disposition had changed—my sadness replaced by a hopefulness that was almost joyful. I put my hand on the chain on Nama's neck. "It's coming off, Nama. You'll be coming with me very soon." She listened as she chewed her papaya. After we gave food to all the monkeys and distributed water to everyone, Estelle and I were eager to discuss details of the pending relocation with the hotel manager we had come to know.

When we asked for the manager in the hotel office, we learned that he wasn't there. In fact, he was no longer the manager. A younger sibling, drunk on his newfound authority and significantly more belligerent than his brother had been, was now in the role of manager. We met with him at a corner table in the Luna Park restaurant. From where we sat I could see Nama, at the limit of her chain, still chewing the food we had left and cocking her head to try to see us. The manager kept us waiting while he gave instructions that had nothing to do with us to two restaurant employees, then summoned a groundskeeper to give irrelevant—from our point of view—directions to him as well. Finally, with an air of casual indifference, he turned his attention back to us. Calmly, he informed us that we would not be allowed to take Dorothy and Nama. Unconcerned that he was breaking his brother's verbal contract with us, he said simply, "My brother is no longer in charge of the hotel or the chimpanzees. My father is making the decisions." With forced composure, Estelle translated all that the infuriating, disdainful young man said, although I had pretty much understood it.

"Is your father here?" I asked. He was. Estelle and I hardly spoke as we sat alone and waited for the manager to find out whether the elder decision maker would grant us an audience. He would.

With an air of arrogance, our informer led us out of the restaurant and around back to a private patio where his father sat. The patriarch wore a navy floor-length robe, and I noticed a walking cane leaning against his chair. After we shook hands, he gestured for us to sit in white plastic chairs across from him. We started off with small talk, as is customary in Cameroon. Estelle thanked him for receiving us. She translated my compliment that his hotel was beautiful. (Although I was by now speaking some basic French to my staff, my vocabulary was small, my grammar terrible, and I certainly would not speak French in a situation like this where nuance was critically important.)

In response to Estelle's query about how old Dorothy and Nama were, he told us that he had acquired Dorothy when his oldest son, now forty years old, was a baby. He said they had bought Nama in 1984. As I did the calculations in my head, Estelle went on to explain who we were and what we were doing in Cameroon. So Nama had been here for sixteen years. For some reason, I couldn't bring myself to accept that Dorothy had endured this "life" for forty years. We had independent confirmation, from someone Estelle knew in Yaoundé who had met Dorothy when he visited Luna Park as a child, that she had been chained here for at least twenty-five years. I couldn't bear to think she had been here longer than that.

"Bullshit," I responded when Estelle mentioned it later, without any sound reason for my skepticism. "I'm sure he doesn't remember when she came," I said.

When Estelle finished introducing our mission, the distinguished looking, regally behaving patriarch acknowledged his understanding with a nod. The several moments of silence that followed served as my cue that it was time to get to the substance of the meeting.

Estelle translated as I started off in my most respectful tone. "After your son told us we could take Dorothy and Nama, we prepared a very good place for them where they will be able to move freely, off their chains. Soon they will live with other chimpanzees in a tract of forest we have enclosed with electric fencing. They will move through the forest like they did with their mothers when they were babies. They will be happy."

"My wife and I are accustomed to seeing them here. We will keep them here," he said matter-of-factly, as if that matter were finished. He wasn't accustomed to being challenged. I certainly couldn't leave it there. I couldn't believe he was condemning Dorothy and Nama to continuing hell, and breaking my heart, with such a casual tone. Surely he would see reason, would have some compassionate feelings toward the chimps. I struggled to keep my tone respectful, while I grappled with the realization of what was happening and reacted to it physically with something akin to nausea.

"I understand your sentiment toward them, but this is not a good life for them," I said. "If you care for them, please be kind to them and let us take them to the forest. We have an enclosure that is ready and waiting."

"I will build an enclosure for them here, better than what you have built," he informed us. I had not expected him to say that and was not prepared with a diplomatic reply.

"C'est ne pas possible pour vous." *It's not possible for you.* I blurted my too honest response in very bad French. I should have stuck to English. Probably more than any other exchange between us, these words gave rise to his feeling, later expressed to the Ministry of the Environment and Forestry through his attorney, that the American woman, specifically me and not Estelle, was a racist.

"You think you can do it, but I cannot? Why do you think I cannot accomplish what you can?" The volume of his speech didn't change, but anger drenched his words.

Indeed, I didn't think he could, or would if he could, build an electric enclosure for Dorothy and Nama. Only a few people in the world possessed the expertise required to construct an electric fence that would actually contain chimpanzees. The expensive building materials were not available in Cameroon and had to be imported. Logistical difficulties aside, to believe that this man would go to the trouble and expense of even trying to build a big electric enclosure when he had let Dorothy and Nama live on chains for decades would have been an impossible stretch of my imagination. If he had possessed the slightest sensitivity to their feelings and welfare, or to that of the eight monkeys who were also tethered, he could have extended some effort on their behalf years earlier—even to build a cage.

Now, more than a decade later, I see possible shades of gray in his motivations and more options in my own choices of behavior and words. Perhaps the mere fact that this elderly man of such high status had agreed to meet with two powerless—or so we all thought—Western women reflected some decency that I couldn't see at the time. After all, no one forced him to meet with us. Maybe he had hoped that we would be a resource to provide something better for Dorothy and Nama right there on the manicured grass of Luna Park. A self-serving notion to be sure, but it would have incorporated a better future for the chimpanzees— and possibly the monkeys too. We never got as far as discussing anything like that. I don't know if he had ideas to which he never gave voice, perhaps impeded by my impulsive reply.

I met his angry stare with a suppressed fury that at least equaled his. "Is that your final word? You are refusing to make the humane and decent decision for Dorothy and Nama?"

As Estelle finished translating, he reached for his cane as if to rise.

"Is that your final word?" I insisted. I was drawing battle lines. I had no common ground with this man. I knew he would never agree to let us take Dorothy and Nama, and I would never stop trying to get them out of that place through any means I could find.

"Yes," he answered simply as he met my eyes a final time before standing up. The meeting was over.

Estelle and I stood, and with forced civility that felt ridiculous, we each shook his hand. On the walk back to the car my anger gave way to the deep sadness that spawned it. I didn't want

Dorothy and Nama to see me so sad, and Estelle must have felt the same. Neither of us went to say good-bye.

Back in the car on the drive back to Yaoundé, I finally broke down and cried. Silently, Estelle smoked a cigarette and waited for my crying spell to pass. She understood that any words of comfort and reassurance she could offer would be empty. After a few minutes, I blew my nose and sniffed a couple of times, a signal to Estelle that I was done.

"Let's go by the Ministry of the Environment and Forestry office on the way home," she said. "We'll see what they have to say."

The timing of our visit to the Central Province's Ministry of the Environment and Forestry (MINEF) office in Yaoundé was fortuitous. MINEF's divisional delegate for the Obala region happened to be attending a meeting there when we arrived. Mr. Daniel Essi was just who we needed to see. After we described our new sanctuary in the Mbargue Forest and told him about our experience at Luna Park earlier that day, the delegate told us about his own frustrating experiences with the politically powerful family. Representing MINEF, he had asked the family repeatedly to acquire permits that would authorize them to hold chimpanzees and monkeys at Luna Park. Because the family had acquired Dorothy and Nama before Cameroon's law protecting endangered species was strengthened in 1994, MINEF was willing to give them an exemption to the provision of the law prohibiting private ownership of chimpanzees. They could have paid permit fees, equivalent to $400 for each primate, and kept the chimpanzees and the monkeys. That they refused to pay the fees indicated to Mr. Essi that they thought they were above the law. He felt that they had disrespected MINEF and him personally, and he was angry about it.

He told us he was willing to confiscate Dorothy and Nama, but only if we could arrange a home for all the monkeys so he could seize them at the same time. He could not enforce only part of the law. We could take all or none of the primates. When Estelle and

I walked out of the office, we were not yet jubilant, but the world looked completely different than it had two hours earlier when I was sobbing in the car—it was now full of possibility for Dorothy and Nama. When presented with the necessity of rescuing the monkeys too, I felt ashamed that we had been willing to leave them behind. To do otherwise had not seemed possible before because they weren't protected species. But in any case, we had thought the family was so powerful that the government wouldn't move against them, so we had wasted time trying to get an agreement with the family for Dorothy and Nama.

We reached out to Limbe Wildlife Center (LWC) and to Cameroon Wildlife Aid Fund (CWAF), which operated Yaoundé Zoo, about taking the monkeys, providing them details about the species, age, and gender of each one. Limbe Wildlife Center was willing to take three of the monkeys, including the adult baboon, and Yaoundé Zoo would take five smaller juveniles. Australian Dave Lucas, the new manager at LWC, and Jonathan Kang, the head caregiver there, would come to Obala in their pickup truck to provide technical and logistical support for the confiscation. Then they would transport three monkeys back to LWC. Bibila Tafon, "Dr. Babs," the talented veterinary technician from Yaoundé Zoo, would also come with a caregiver and two vehicles—including one taxi—to fit five monkeys. When the necessary arrangements had been made, Mr. Essi set the date on the second Tuesday of May 2000, when we would all convene in Obala for the operation. I went back to the sanctuary in the interim and returned to Yaoundé with Kenneth in our old red Toyota pickup, which I had bought after selling the Pajero. On the fated Tuesday, Kenneth, Estelle, and I crowded into the single cabin for the drive to Obala. I had gotten up before dawn after hardly sleeping. I was nervous and so hoping that nothing would go wrong.

While Dave, Babs, and their caregivers waited alongside the road in front of Obala's open-air market—the very market where

I had battled my conscience over the dire fate of a doomed crocodile more than a year earlier—Kenneth, Estelle, and I joined Mr.
Essi and three of his forestry officers at the Obala headquarters of
the military police, only a three-minute drive from Luna Park. The
bright hunter-green uniforms of the forestry officers contrasted
sharply with those of olive green worn by the more authoritative
military police, who topped their ensemble with a distinguishing
red cap. Concerned that news of the planned confiscation might
leak out to the politically connected Luna Park family, Mr. Essi
didn't even tell the military police commander in advance what
we would be doing. Instead, he had requested four officers for an
unspecified mission. Only on that warm, sunny morning, con-
vened under the big mango tree outside the commander's office,
did Mr. Essi explain to all seven armed participants that they
would move into Luna Park for a forced seizure of ten primates.

"Your job will be to protect the technicians"—he gestured to
us—"while they load the primates into cages and onto their vehicles. They will need some time to do their work, and you should
be prepared for resistance." The military officers, each holding
automatic rifles, exchanged tentative glances with one another.
I wondered if they knew the proprietors of Luna Park personally, since they all lived in this small town, and might be uncomfortable with the operation, but I thought they looked prepared,
more or less, to move forward. They really didn't have any choice
since the commander had issued their orders.

"Any questions?" Mr. Essi asked. None were voiced.

During the three-minute drive to Luna Park Hotel, the atmosphere in our pickup was tense. No one said a word. Our caravan
of seven vehicles, with Mr. Essi in the lead, pulled into the long
driveway. We parked our vehicles along the side and filed out
onto the grass. The officers spread out, positioning themselves at
the periphery of our work area, each holding his weapon diagonally and very visibly in front of his torso. After we offloaded our

cages, Estelle led Dave and Babs to the seven monkeys tied on the veranda. They had seen the adult baboon chained near the driveway on our way in. Thereafter, each team proceeded to work independently and quickly, while Mr. Essi went to the office to serve papers.

I was available if Dave's team or Babs's team needed me, but they didn't. They darted the adult baboon and three older juvenile monkeys with anesthesia and were able to carry and place the four youngest into cages without anesthesia.

Estelle, Kenneth, and I focused on the chimpanzees, first carrying one of the transport cages into Nama's area. She expressed interest in the cage immediately, pushing up the sliding guillotine-type door herself. "Nama may go in without anesthesia," Estelle said hopefully, while helping Nama lift the door up far enough for a curious chimp to crawl through the opening. When Nama was completely inside, Estelle closed the door quickly and threw herself across the top of the cage, holding the door closed with her body weight. Her weight would not have been enough to contain most adult chimpanzees, and even frail Nama might have gotten out if she had tried. Fortunately, Nama was concerned, but she didn't panic, and her efforts to push the door open again were halfhearted. She watched with great interest while I used our big bolt cutters to sever the chain tying her to the concrete slab. Estelle opened the door just enough for me to push the cut end of the chain into the cage with Nama, so we could completely close the cage door and slide a lock through the latch. It had been amazingly easy! Nama had developed a degree of trust in us, the only friends she had had in many years I reckoned, and I knew she was desperate for any change in how she was living. Once we were safe at Sanaga-Yong Center, I would need to anesthetize her to cut the tight chain from her neck. I harbored no illusions that Nama trusted us enough to let us approach her neck with those big bolt cutters—she likely would have confiscated them for herself.

Unfortunately, Dorothy had no intention of entering a cage willingly. Under different circumstances, she might have allowed us to give her an injection of anesthesia with a handheld syringe, which is much less painful than a dart, but not this day when she was agitated by all the strange activity around her. She was wary of the syringe and wouldn't have it anywhere near her. I had to blow a dart of anesthesia into her big thigh. When it hit her, she screamed and grimaced, and I hated that my sweet friend thought I was betraying her. Fortunately, it was over quickly.

When Dave and Dr. Babs had collected their monkeys, they helped us carry the cages with Dorothy and Nama to our truck and load them in the back. Within forty-five minutes of our arrival at Luna Park, the chimpanzees and all the monkeys were loaded into the four vehicles that would take them to better lives. Just as we were preparing to leave, Estelle went to assure the concerned-looking employees, who had congregated on the steps of the restaurant, that all the primates were going to places where they would be happier. Standing at the bottom of the short staircase, she spotted two sick, almost featherless parrots in a tiny cage in the back corner of the open-air restaurant. We had not seen them there before. "I wonder if we can take two sick parrots from the restaurant?" she said when she came back to the truck where I was watching and comforting Dorothy as she began to wake up from the light anesthesia.

"Ask Mr. Essi if we can take them," I suggested.

I watched Estelle approach Mr. Essi's car, where he was waiting for us to finish and leave. The two spoke through his open window for less than a minute, before she returned to me.

"They don't have permits for the parrots, either," she said. "We need to take them."

I knew that Chris Mitchell had just completed a small aviary at Yaoundé Zoo. Dr. Babs agreed to take the parrots there until they might be released one day, and he sent the caregiver working with him into the restaurant to bring the miserable birds out.

We had lined up the cages of Dorothy and Nama side by side, front to back, in the bed of the pickup so that there was space for me to sit beside them. I would ride to the sanctuary back here with the chimps, which was only slightly less comfortable than squeezing onto the single-cabin truck seat with Estelle and Kenneth would have been. This was especially true since we had another passenger in the cabin. The past January I had assisted MINEF authorities in the town of Kribi confiscate four-year-old chimpanzee Caroline from a small cage at a restaurant. Lacking a quarantine facility or a nursery at our sanctuary, I had arranged with Dave to keep her at Limbe Wildlife Center for three months, and it had stretched into five. On this day of the confiscation, he had brought her with him from Limbe to hand over to us.

Our red pickup had been last in line heading into Luna Park; now heading out we were at the head of the caravan. As we neared the end of the single-lane driveway, we were forced to stop when a big black Mercedes pulled in. I stood up in the back of the truck as the driver of the Mercedes quickly maneuvered it sideways across the road, blocking our exit. Luna Park's proprietor, the elder family patriarch, stepped out of the backseat in a yellow floor-length robe and planted his feet and cane defiantly.

Mr. Essi walked up from behind us to manage the problem. "Sir, this is an official action by the Government of Cameroon. We have seized the animals you were holding illegally. Please instruct your driver to move the car."

"No one is taking my animals anywhere," our nemesis responded with conviction.

"It's too late for that, sir. Your animals are being distributed among the three MINEF primate sanctuaries, and you can visit them there." Mr. Essi turned and instructed the military police, who had also walked forward, to clear the road. With that, he proceeded back to his own car, leaving the police officers to handle the obstacle.

The patriarch got back into his car. After a minute or so, with no indication that the car would move anytime soon, one of the police officers instructed us to drive onto the grass and around the front of the car. When Kenneth tried to do it, the driver of the Mercedes also pulled forward onto the grass to block us. As Kenneth returned to our original position on the road, the driver also backed up to straddle the middle of the road once again. I was sensing that the military police officers weren't eager for a physical confrontation with the car occupants; they probably knew one another. Their next strategy was for us to try driving on the grass around the backside of the car. Doing as he was told, Kenneth steered our pickup hard to the right and inched onto the grass as if to go around the back of the car. No one, certainly no one in our pickup, was the slightest bit surprised when the Mercedes driver backed up onto the grass to block us. With both vehicles back in their original positions, we watched the military police officers discuss their options among themselves. Finally, one of the officers moved quickly to open the driver's door—fortunately the window was down so the driver couldn't lock it—and another pulled him out of the Mercedes. Now we could drive around the car.

The patriarch got out of his car and pointed his finger at me menacingly. As our truck rolled past him, I met his hate-filled glare without embarrassment. Turning my face to the road ahead that would carry Dorothy and Nama to their new life, I couldn't have been happier.

ten

WHO'S THE BOSS?

S oon after their arrival Jacky and Pepe started fighting for
dominance, and it continued throughout the months before
Dorothy and Nama arrived. Becky enjoyed loving relationships with both males, and she was neutral in their extended contest. Because she and Pepe had grown up together since they were
babies, they had a brother/sister relationship that was not overtly
sexual. Jacky chose masturbation over copulation; for decades
he had known nothing else. Nonetheless, the fights between the
males were more frequent when Becky had her genital swelling—a descriptive term for the monthly vulvar engorgement that
occurs in female chimpanzees before and during ovulation.

The violence upset Becky terribly, and she tried her best to keep
the peace and to break up fights when they happened. When she
sensed that Jacky or Pepe was in a menacing mood, she tried to
calm them with grooming or distract them with her sexuality
if she was in swelling or position her body between theirs. She
screamed and slapped and tried to intervene during the heat of
their battles, but she did not seem to favor one over the other.
She spent time engaging socially—grooming and hugging—with

both of them. I knew from reading that group support can be very important in establishing and maintaining the dominance hierarchy among chimpanzees, but in the circumstances I had read about, an established dominant male was forced to fend off challengers. The length of a male's reign and the success or failure of his challengers was highly influenced by who was on their sides. Between Jacky and Pepe nothing was established, neither had acknowledged the dominance of the other, and the one neutral female in the small social group was not hastening an outcome. I briefly considered keeping them apart until we had a larger community of chimpanzees. I certainly would have done that if it had seemed that one needed protection from the other, but when Pepe and Jacky fought, they were both willing combatants. Neither was submitting to the other, which is all it would have taken to stop the fighting, and I kept thinking they would sort it out sooner or later. After each fight they reached a temporary truce, which usually started with Pepe making a tentative approach to groom Jacky, but sometimes it was the other way around. According to the caregivers, Pepe started more of the fights than Jacky did. In times of peace, they engaged in mutual grooming sessions and even hugged when they were excited about a favorite food or frightened by an unusual occurrence, like an airplane overhead. Sometimes weeks could pass without bloodshed, but when tension became too thick, it was vented through loud and frightening fights that served to temporarily restore calm.

The battles exacted a toll on both of them, but their wounds were mild compared to what I would see in some chimpanzees in years to come. I cleaned gaping bite wounds on Pepe's back and arms, but nothing severe on his face. At first, Jacky didn't seek my attention for his wounds the way Pepe did. I wanted to provide medical care to both chimps, although I was still wary of Jacky. When I called to him from the side of the cage, usually somewhat tentatively, he pretended to ignore me, although I sometimes knew he was watching me tend to Pepe's wounds. I soon realized that by taking care of Pepe, but not Jacky, after their fights, I was playing a role in the social dynamic—giving support to Pepe in the unsettled conflict. I didn't know how significant my role was, but in any case, I really did believe that Pepe would be a better leader of this little group that would soon expand. Pepe was the sane one, after all.

However, after about three months at the sanctuary, Jacky's stereotypical self-abuse had stopped almost completely, and he seemed like a different chimpanzee. One afternoon he surprised me by initiating a change in his relationship with me. I was just outside the cage grooming Becky through the bars, and Pepe was on the opposite side of the cage visiting with caregiver Akono. Jacky suddenly walked up next to where Becky sat across from me, and she considerately scooted over to make room for him. He was only two feet away from me, and although the cage wall separated us, the holes in the metal latticework of that first cage were large enough for his arms to fit through. Easily within range of his arms, I leaned back away from the cage reflexively, but I stayed seated. As Jacky faced me, he kept his eyes down, and I had no idea about his intention until the moment he turned to push the right side of his big black back against the cage wall. He was showing me the deep gash extending into his muscle under his right shoulder blade. I had seen the wound inflicted by Pepe during their last altercation two days earlier, but I hadn't held any hope of treating it.

I could hardly believe that Jacky was asking for my attention to the wound. He was choosing to trust me. It should have been natural for me to trust him in return, at least in this particular circumstance where he was asking for my help, but having seen how he had hurt people, I was still nervous. My heart raced as I examined the infected wound. As my shaking fingers touched the skin around it, I made rhythmic clacking, smacking noises with my mouth to reassure him. When Jacky relaxed into a comfortable sitting position, keeping his back to the cage wall, I relaxed a bit too. Worried that he would leave, I hesitated before walking to collect the disinfectant from the nearby table, but he waited patiently for me to return and allowed me to flush the debris and pus off the exposed muscle in the deep laceration. After that first occasion, Jacky solicited and received my attention to his wounds on his terms, when he decided he needed it.

Finally, in April 2000, about a month before we brought Dorothy and Nama to the sanctuary, we released Jacky, Pepe, and Becky into their five acres of forest. The electric fence around the tract of forest consisted of thirty-two wires, each carrying eight thousand volts. Each of the three chimpanzees had reached through their cage and taken a nasty shock from the hot wires before we released them into the enclosure. I had designed the fence with some of the wires passing close to the cage so the chimpanzees would learn about its unpleasantness from the beginning. Because the wire fence itself was not a strong physical barrier and the electricity running through its wires could be grounded out by a falling tree or limb, it was important for the fence to be a strong psychological barrier. This aspect of the process, allowing them to naively touch the fence, was mean, but using electric fencing was the only way we could afford to give the chimpanzees their piece of the forest. By the time we opened the sliding door to let them go outside, they knew that touching the fence was a very bad idea.

When caregiver Assou pulled open the sliding door for the first time, the chimpanzees didn't know what to do at first. They rushed to huddle at the open door and then just peeped through at the expansive forest on the other side. Becky looked up at me quizzically like she thought it was a trick. "It's okay, Beck. Go outside." We had cut narrow crisscrossing trails that traversed the expanse of the lush forest, and we had cleared about two yards along the fence line so they could easily walk around the perimeter of their new territory. It was all waiting for them, and I could hardly wait to see them enjoy it. I advanced along the fence line and called to them over and over while they continued to think it over for several minutes. Finally, Jacky marched out bravely through the sliding door, but he only walked about five steps toward the forest before he turned to run back into the cage. In the end, Becky took the lead. Clearly afraid of the electric fence, she sat in the doorway and inched her head through it very slowly to look up at the hot wires over her head. When she was sure that there really was an open path from where she sat to the forest, she made a run for it. By the time Jacky and then Pepe had followed her out, she was already climbing a tall tree. I started walking along my side of the fence line and called them to follow. The immediate pitter-patter of Pepe's feet on the firm red dirt told me that he was with me before I could turn my head to see him. Jacky and Becky followed after him, and we walked—side by side—them in single file on one side of the fence, me just a few feet away on the other—around the enclosure for the first of many, many times.

After lengthy consideration I had decided that the chimpanzees would spend their days in the forest, and we would try to bring them into the satellite cage to sleep at night. While I wanted them to have a natural life, I had good reasons for wanting them to sleep inside. First, if a storm came and dropped a tree or branch that grounded out the fence while we humans were sleeping, I didn't want to worry about where the chimps were in the morning. Second, getting the chimpanzees to sleep inside the cage would spare the forest of their enclosure. Night nests can be destructive to trees, and if the chimps nested outside every night, they would eventually break down their tract of forest. This would be especially true when we enlarged the group. And third, bringing our beloved chimpanzees into the cage every evening would allow us to know if anyone was sick or injured.

Long before we finished the electric enclosure and let Jacky, Pepe, and Becky into it, we began beating our drum at each mealtime. What we called our "drum" was actually a five-gallon plastic container that made a loud noise when our caregivers hit it with a stick. They tapped out the same rat-a-tat-tat pattern every day, just before each meal, so the chimps understood it signaled the serving of food. Now that they were going outside, we used it successfully to call them back to the cage for meals. When they came into the cage for dinner at the end of each day, we closed the sliding door until morning, and no one among the three ever objected. They seemed to be perfectly comfortable with the schedule.

I soon learned my own healthy respect of the electric fence the hard way. One day when I was kneeling close to the fence taking photos of the three chimpanzees, I made the mistake of touching one of the hot wires with both my knees. I was knocked back on my behind and I must have yelled, although with the force of the shock, which sounded like a gun going off in my head, I couldn't tell if I made any noise. When I caught my breath, I looked out into the empathetic, concerned faces of Jacky, Pepe, and Becky. They had run up to the fence and were now watching me very closely. Of course they understood what had happened to me. "I'm okay, guys." I stood up to demonstrate. I *was* fine, but my ears were ringing and I needed to take a break for a while.

I hoped none of these chimps I loved would ever touch the fence again. I hoped that their new home would be rich and wonderful and that they would never want to leave it—would never risk getting shocked. The relatively vast territory within the enclosure offered so much more to explore and enjoy—leaves, bark, insects, an expansive view from high in the canopy—than they had ever known. I had left my own home and dramatically changed my life in order to give it to them, and when I was with them here I didn't regret it in the slightest.

eleven

FATEFUL ALLIANCES

Dorothy and Nama spent their first three months at the sanctuary together in a satellite cage with two chambers. Only one chamber connected to the enclosure where Pepe, Jacky, and Becky spent their days. We had tested Dorothy and Nama for tuberculosis before bringing them from Luna Park, but I wanted to treat them thoroughly for parasites and keep them separated from the other chimpanzees past the incubation periods of most infectious diseases. The standard medical quarantine period for primates in sanctuaries was three months. Ours was a loose form of quarantine. Jacky, Pepe, and Becky could emerge from the forest to see Dorothy and Nama and to vocalize with them whenever they felt like it, but they couldn't get too close to them. It was a good beginning for a gradual social integration.

Initially, the vocal overtures of Jacky, Pepe, and Becky were threatening barks. They frightened Dorothy terribly. She screamed and took comfort in the arms of much smaller Nama, who from the beginning seemed more curious than afraid. When we had first released the two long-suffering females into the cage together, they had stayed several feet apart for a few minutes. I had expected them to fly into each other's arms and was surprised when they didn't, but I concluded that they just weren't accustomed to getting comfort from one another that way. When Nama finally made an overture toward Dorothy, the latter welcomed her with open arms, and then they embraced over and over. During this transition period at the sanctuary, when they were in the cage alone together, Dorothy and Nama cemented a lasting and loyal friendship that would never waver.

The thirteen feet of vertical space in the satellite cage, with platforms at various levels, provided opportunities for Dorothy and Nama to climb and strengthen the muscles in their arms. Nama took full advantage and grew stronger by the day, but Dorothy didn't climb much. I thought the chain had grounded her for too many decades. Nonetheless, as the days passed Dorothy seemed happy to groom with Nama and to interact with her caregivers, the volunteers, and me from where we would sit just outside the cage. I tried to provide as much enrichment as possible. They had been traumatized by a decades-long nightmare, and though this new environment afforded them the freedom of movement they hadn't known since they were juveniles, and allowed them to engage as often as they liked in the most important chimpanzee social activities of grooming and hugging, they were still in a cage. It was much less than they deserved, and I delighted in making them happy in any way I could.

One cool morning when it had rained the night before, Dorothy and Nama were both shivering when we delivered their breakfast at seven o'clock. Seeing that they were cold, I rushed back to the camp to make them some hot tea. Their breakfast that morning had included a half baguette each of French bread, simply called long bread by the locals. Roger Odier had brought a big bag of it from Bélabo. When I handed them their plastic cups of tea, which barely fit through the holes in the cage wall, a half hour after the breakfast had been given, Nama took a sip and then immediately reached around behind her to retrieve the uneaten half of long bread she had left on the cage floor. Without a moment's hesitation—she knew exactly what she was doing—she dipped the end of the big piece of bread into her cup of tea and bit off the soaked part, grunting with delight as she enjoyed the special treat. While her mouth was full and chewing, her eyes found mine for a brief glance of gratitude before she turned her full attention to the sweet indulgence.

Dipping bread in coffee is customary among many Cameroonians, a breakfast tradition shared with them by French colonizers. As a juvenile, Nama must have lived in a house among people, either French or African, who had dipped their bread—possibly the family of Luna Park proprietors, which had kept her cruelly chained for so many years. Someone must have shared the practice with her when she was young, and she still remembered it after so many years.

During those first months when Dorothy and Nama were in the satellite cage, I indulged them with the tea and bread breakfast after each of our weekly trips to the small bakery in Bélabo. At first, Dorothy ate her bread dry and sipped her tea separately, but after a couple of weeks, she too was enjoying her bread soaked with tea. She, however, used a different method. Instead of dunking and biting from the end of a big piece of bread as Nama did, Dorothy pinched off small pieces, dipped them into her cup of tea, and plopped the whole morsels into her mouth, chewing and swallowing them one at a time. As a captured infant, decades earlier, had Dorothy too watched humans eat this way? Had she herself partaken of tea or coffee with bread? It could have explained her different dunking method. Or perhaps she simply got the idea from Nama and adapted her own way of bread dipping. Of course, I had no way of knowing, but I wondered about it as I watched them sitting quietly together, using different dunking methods to enjoy their bread with tea.

Another clue that Nama had lived in a house was her propensity for cleaning. One morning I arrived at the cage to discover that Nama was holding a washcloth-size piece of one of the big blankets I had left for her and Dorothy. I didn't see whether she had torn it off intentionally with a purpose in mind, but she was guarding it carefully. Later in the morning when the caregivers poured water and soap on the cement of the cage to perform their daily cleaning, Nama used her cloth to help them scrub, periodically wringing soapy water from it. Our staff used brushes and brooms to clean the cages, so Nama hadn't learned to use cloth from them. She had seen that somewhere else, and it had been many years earlier that she had seen it. No one had been cleaning the dirt where she was chained for sixteen years.

After three months of visual contact, with the period of medical quarantine over, I allowed Dorothy and Nama to interact with Jacky, Pepe, and Becky through the side of the cage that connected to the forested enclosure. Nama made friends with the two males quickly, and with Becky a little more slowly. In the early days of the integration, her relationship with Pepe was playful. She often lay on her back, up against the cage wall, squirming and laughing heartily as Pepe reached through the cage with both hands to gently poke and pinch her belly and the backs of her thighs. In those moments, Nama was a kid again. With Jacky, she groomed more often than played. Nama and Becky sat calmly within each other's proximity, sometimes grooming, often just sitting.

Unfortunately, Dorothy was socially awkward. Like Nama, she had not been part of a chimpanzee society since she was an infant, but with Dorothy it had been decades earlier. In addition, Dorothy may have felt disabled by her obesity and general lack of physical prowess—conditions imposed on her by a high-fat diet of palm nuts and the chain that prohibited all exercise. Whatever the reasons, Dorothy didn't know how to act with other chimpanzees, and making friends was hard for her. She took much longer than Nama to approach the cage interface where the other chimps beckoned. Although she developed friendly relations with Jacky and Pepe, her tentative shyness irritated Becky, who often barked and threw dirt at her. Within the protected cage, Dorothy could keep her distance from Becky, but our goal was to have them all in the forest together.

Fearing that we might never see the kind of positive affiliation between Becky and Dorothy that would lead me to feel 100 percent comfortable about putting the five chimps together, I decided to move forward with the physical integration one step at a time. I felt sure that Nama was ready to go out in the forest with the others, and I didn't want to keep Dorothy inside alone.

The hot wires of the electric fence ran close to their cage, and Nama had touched it. I never saw Dorothy touch it, and neither did her caregivers. She may have touched it when we weren't around, or she could have taken Nama's advice that she shouldn't.

I started by allowing Nama and Dorothy into the forested enclosure with Pepe, keeping Jacky and Becky in their satellite cage for the afternoon. It was easy to get them in the cage, because we normally beat the drum to call them all for food twice during the day and to come in for dinner at the end of the day. When Jacky and Becky came in for their two o'clock snack, we locked the sliding door behind them. They didn't mind too much since they had been out all morning. This choice of leading the integration effort with Pepe reflected my bias that he would be the better leader of the group. When we opened the sliding door of Dorothy and Nama's cage, Pepe squeezed past Nama to enter the cage as she eagerly ran out of it. Immediately he turned and followed her out, staying playfully on her heels as she trotted around in front of the cage, laughing gleefully. After these few moments of elation, Nama realized that Dorothy was afraid to come out of the cage. Reaching out her arm, palm up, she beckoned her less confident friend. Noisily and with a grimace of fear, Dorothy responded.

FATEFUL ALLIANCES

149

Like she was taking a desperate, fearful plunge into a pool, she rushed through the door screaming into Nama's, and then Pepe's, waiting arms. Pepe was gentle and kind that first day.

With Pepe in the lead, the three started down the foliage-free dirt path that ran along the fence line. Jacky, Pepe, and Becky had maintained the footpath with their regular patrols of the fence perimeter, so it seemed natural that Pepe wanted to take Dorothy and Nama along the path. I followed along on the opposite side of the fence. Dorothy hadn't walked more than a few feet at a time in decades, and when she fell behind, panting and sweating, it was Pepe who came back to her. From Dorothy's right side, Pepe wrapped his big, muscular arm around her back and under her left arm to pull her along, encouraging her to continue. With his help, Dorothy made it the half a mile around the enclosure periphery, past the satellite cage from where Jacky and Becky watched quietly but with great interest, and back to snacks and water in the cage. Afterward, while Dorothy rested in the cage, Nama and Pepe entered the forest, out of her sight and mine, for several hours. While Dorothy watched eagerly for Nama's return, she didn't seem distressed by her absence. At the five o'clock dinner hour we beat our plastic "drum" to let Pepe and Nama know that food was being served at the cages. Hearing the familiar beckon, Nama returned to a happy welcome from Dorothy, while Pepe returned to the cage he shared with Jacky and Becky. All in all, I felt that Dorothy's and Nama's first day in the forested enclosure with Pepe had gone very well.

The following day, I kept Becky and Pepe inside, while Jacky went into the enclosure with Nama and Dorothy. This day was a success, too. After Dorothy slowly made her way around the perimeter of the enclosure, she collected a stick from the edge of the forest and used it to groom herself while she relaxed in the shade just outside the cage. I was happy that, in contrast to the day before, she chose a resting spot outside of the cage. Like she had with Pepe the day before, Nama spent time in the forest with Jacky, and they returned calmly to their separate cages for dinner.

Nervously, on the third day, I allowed all five chimpanzees in the enclosure together. Estelle had arrived on the train the evening before for a short visit, and she encouraged me to move forward. As usual, I was glad she was there.

This first day that all five chimpanzees were together gave birth to an amazing and unexpected alliance that would establish the hierarchy of the group for the next decade. Jacky, Pepe, and Becky were congregated outside Dorothy and Nama's cage when we opened the door. Coaxed by Nama pulling on her arm, Dorothy passed through the door, scared but trusting, and followed Nama several yards into the enclosure. With five chimpanzees, the social dynamic was more complicated. With my heart pounding like I had just run a mile, I watched with Estelle, Kenneth, and the caregivers as all five chimpanzees sat around the front of the cage with a nervous energy in the air. While Becky wasn't overtly friendly and welcoming to either Nama or Dorothy, she wasn't behaving aggressively either. I had just released a sigh of relief about Becky's behavior when Pepe surprised me. Hairs on end, he ran stomping past Dorothy and slapped her hard on her relatively frail back, the sound resonating loudly against her chest wall. She screamed in confusion and looked to me for help. Equally surprised, I had no good options for helping her, other than to get her back in the cage and close the door. Estelle and I called for her to come, and she tried but only got as far as the outside cage wall, against which she cowered in terror as Pepe charged at her again, puffed up and terrifying.

"Pepe, no!" Estelle and I both shouted, although I doubted whether our scolding would have any effect. Fortunately, Dorothy didn't need to rely on us. Nama intercepted Pepe, and the events of the next few seconds happened so fast that it was impossible to comprehend them as rapidly as they unfolded. About half

Pepe's size, Nama plowed into him from the side, catching him midstride. For several moments I couldn't discern who was who in the screaming, blurry mass of brawling chimpanzees, much less who was getting the best of whom. Jacky and Becky were vocalizing from the sidelines, but they didn't enter the battle physically, and I couldn't tell who they were supporting in this initial stage of Nama's fight to protect Dorothy. I was terrified for Nama, but when she and Pepe broke apart, it was he who fled into the forest, screaming. Nama sat near Dorothy panting from exertion, and while she caught her breath, she never took her wary eyes from the forest trail Pepe had entered.

Five minutes later, Pepe returned, exiting the forest on a different trail, with his eyes trained again on Dorothy, who was still sitting with her back up against the cage wall. I had called to her over and over after Pepe had left, but she had chosen not to enter the cage. She was trusting Nama to stand their ground. As Pepe approached slowly but menacingly, Nama paced back and forth in front of Dorothy, aware of Pepe's every move, like a dedicated, lionhearted sentry. She probably couldn't have ultimately prevailed against Pepe, given the disparity in their sizes, but it was crystal clear that she was prepared to fight again.

Not only was I terrified for both Dorothy and Nama, I was upset and disappointed on a personal level that Pepe was behaving so badly. Two days earlier, his kindness toward Dorothy brought tears to my eyes. Now he was trying to assume dominance by cruelly bullying her, the weakest in the group. I knew that male chimpanzees were all about power and dominance and that they commonly bullied weaker males to make themselves appear stronger. There were no weaker males in Pepe's group—only Dorothy, who posed no threat to him. He was transferring his aggression from Jacky, the real target, onto someone safer—using Dorothy in a display to show Jacky he was dominant. Reflecting in hindsight, it made me think Jacky had already gained an edge in their battle for dominance. Pepe was acting more like a challenger and probably had been for some time. With my limited experience I just hadn't recognized it, and perhaps I was blinded by my bias.

Of course, chimpanzees aren't the only great apes to use cruel and unattractive bullying tactics. Studies demonstrating the universality of bullying in human cultures and the frequency with which it occurs in nonhuman primate societies suggest that its evolutionary roots stem back to our common ancestor with chimpanzees, and even much earlier.* While I knew that Pepe's behavior was typical and that I shouldn't be disappointed with him for behaving like a chimpanzee, I wondered if I had exacerbated the ugly tendency with the way I chose to proceed with the integration—introducing him to the females first, and then Jacky. In any case, Pepe's strategy was not a winning one.

Just at that moment in the dangerous confrontation between Nama and Pepe, Jacky made a decision that would have far-reaching consequences. He took up a position beside Nama, just in front of Dorothy. With only a quick glance in his direction, Nama understood immediately that in Jacky she had an ally. Almost instantly, not waiting for Pepe to charge again, Nama and Jacky took the offensive. Together they charged at Pepe with overwhelming force. They chased him into the forest screaming, and he didn't dare come back for more. Pepe never bullied Dorothy again, and with Nama at his side, older and smaller Jacky would become the definitive leader of the group. Together Jacky and Nama would maintain peace in their community of chimpanzees, which we would rapidly expand with the addition of young orphans, for the next ten years.

*Hogan M. Sherrow, Ph.D., "The Origins of Bullying," guest blog on *Scientific American*, December 15, 2011.

Only a few days after the fateful introduction in the forest, Nama and Dorothy began following Jacky to the cage he shared with Pepe and Becky, making it clear that they wanted to eat and sleep there, so I let them. While Dorothy might not have trusted Pepe, and she didn't ask for his reassurance the way she did with Jacky and Nama, she didn't hold a grudge against him, either. After that frightening day of turmoil, they began to associate peaceably with mutual grooming. Unfortunately, it was Dorothy's relationship with Becky that would be more problematic in the longer term.

Nama and Jacky couldn't solve Dorothy's distressing social problems with Becky as easily as they had protected her from Pepe. Becky was much bigger than Nama, and just as dominant, loved by both Jacky and Pepe. I even wondered if competition with Nama was why Becky chose to bully Dorothy. In any case, Nama must have known to pick her battles carefully. She sat on the sidelines as Becky bullied and tormented Dorothy, humiliating rather than causing physical harm. I quit giving cups of tea when I saw Becky pouring it over the top of poor Dorothy's head, the latter screaming in anguish as the tea dripped down her face. As Becky spat and threw dirt on her, Dorothy sometimes reached out her hand to me, grimacing helplessly and begging me for help. Unfortunately, as much as I wanted to help her, I couldn't solve the problem with Becky either. My angry reprimands went unheeded.

Once, when Becky went too far, I saw Nama draw the line. After hearing Dorothy screaming near the satellite cage, I arrived on the scene at the same time Nama emerged from the forest to discover the cause of Dorothy's emotional plea for help. Becky was hitting her on the top of her head and shoulders with a sturdy stick, three feet long and a half-inch thick. Dorothy was seemingly powerless to stop the abuse, but Nama wasn't. With little drama, Nama quietly interposed herself between Dorothy and Becky. Facing Becky,

Nama wrapped her own hand around the stick and tried to pull it from Becky's hand. During several seconds of intense eye contact between these two strong females, Becky resisted, hanging on stubbornly to the stick and pulling back. I didn't breathe, waiting to see who would win this war of wills. When Becky finally released the stick, the tension dissipated instantly. Nama scooted away and, almost casually, tossed the stick toward the forest. It was still in sight, so Becky could have picked it up again, but she didn't. Nama had made her point, strongly and effectively, without aggression. It was the only time I saw Becky use a stick to hurt Dorothy.

As the months passed, Becky paid less attention to Dorothy. She lost interest in tormenting her, and in fact generally ignored her. Dorothy spent most of her days alone on the ground near the edge of the forest, while Nama and the other chimpanzees enjoyed the forest and strengthened their mutual bonds.

During Sanaga-Yong Center's first year of operation, while we followed with intense interest the adult dramas unfolding in one area of the forest, we opened a nursery in another. I had envisioned "my" sanctuary with a small infrastructure to provide for adult chimpanzees—for the long-suffering captives I had befriended at hotels. I would leave it to Limbe Wildlife Center and Cameroon Wildlife Aid Fund to take care of orphaned infants. I would direct the small "adults-only" sanctuary while I spearheaded campaigns to stop the killing and orphaning of free-living chimpanzees. I was already an activist at heart, and coming to know and love the adult chimpanzee orphans had made the conservation issues personal for me. However, my vision for a small sanctuary wasn't realistic in the context of the country I had chosen. Cameroon's thriving illegal trade in ape meat was orphaning dozens, if not hundreds, of chimpanzee infants every year. Although only a small percentage was ever rescued, the other sanctuaries were filling up fast.

My vision evolved soon after the manager of a logging company stopped by our camp with information about a sad, sick baby chimpanzee in a village sixty miles from us. I was away buying supplies in Yaoundé, so the concerned logger gave Kenneth money for gas to go pick up the baby. Kenneth knew that under no circumstances should he buy the baby chimp. It would fuel the trade and work against our mission. He didn't have any money anyway, so he wasn't tempted. Using only his powers of diplomacy and persuasion, Kenneth managed to drive away from the village with the little chimpanzee. When I returned to camp, I first hoped that we could take the two-year-old boy, who Kenneth and volunteer Roberta Sandoli had named Bikol, to one of the other sanctuaries in Cameroon. Only after learning that neither of them had room for him at the time did I let myself fall in love.

Soon afterward, the brother of a hunter brought six-month-old baby Gabby, thinking he could sell him to us. Kenneth threatened the man with arrest by the military police and sent him on his way without Gabby. The incident, which made me fearful that poachers harboring the idea of selling to us might specifically target chimpanzees with babies, inspired us to publicize in the surrounding villages and in Bélabo that we would never buy chimpanzees, and that people who tried to sell them could be arrested. The first prosecution of a chimpanzee dealer in Cameroon would come six years later, after Israeli national Ofir Drori's Last Great Ape Organization would start collaborating with the Cameroon government on wildlife law enforcement.

In February 2000, when Greg Rossell and Anita Phillips arrived to build our electric enclosure for the adult group, we were caring for babies Bikol and Gabby in our camp. The talented builders found time to contribute in huge measure to our professionalism by building a much-needed nursery complex, which consisted of a wooden sleeping house and a smaller electric enclosure.

By the end of our first year in operation, our juvenile population had grown to seven, ranging in age from one to four years. My vision for a small sanctuary for adult chimpanzees was gone. In the years to come, the majority of the chimpanzees we would rescue would be young infants for whom we would be surrogate mothers until we could integrate them with older chimpanzees. We would eventually integrate many into Jacky's group.

The staff, volunteers, and I alternated taking the babies for long daily walks along a complex network of forest trails. During our excursions, the older babies climbed high in the canopy, while the younger ones climbed lower and played on the ground around us, sometimes napping sweetly on our laps when we rested. We human surrogates were the babies' authority figures and their protectors. We provided love and hugs, but also harsh scolding when it was required to intervene in skirmishes or protect ourselves from naughty behavior, like hair pulling. Because we loved them and they needed all we provided, they were upset if we were angry with them. A reprimand was usually followed by whimpers of apology and quick reconciliation. In the forest, whenever we stood to go, all the babies raced toward us, flying down tree trunks, competing for one of the prized riding spots in our arms or on our backs or clinging to our legs. When I was alone with them, on days we were shorthanded, some would need to simply grab a fistful of my shirttail, or even walk a few steps ahead on the trail. One thing is sure, none of the babies intended to be left alone. They were needy and loving, and their social dynamic was relatively uncomplicated.

twelve

CHALLENGES ON MY SIDE
OF THE FENCE

While I enjoyed the babies and fretted over Dorothy's struggle to find her comfortable place in the small group of adult chimpanzees, I worked to keep harmonious relationships with the people of the village amid a confusing and evolving political and social backdrop. An important occurrence was the enthroning of a new traditional chief in the Mbargue Forest village of Mbinang, about three and a half miles from our camp. All our dealings had thus far been with Chief Gaspard of Bikol 1, but the traditional territory of Mbinang, I soon learned, encompassed the two Bikols and five other small villages. We had first arrived in the Mbargue Forest soon after the longtime traditional chief of Mbinang had died, and there followed a lengthy dispute over who would be the succeeding chief. I might say that the confusion had left a power vacuum in which the leaders of the small villages had been free to assume autonomous rule, but actually, as I understood it, neither the chief nor the people of Mbinang had used this part of the forest for hunting or farming or paid any attention to it whatsoever for many years before. The local population didn't see fit to mention to me that Mbinang politics might affect us, and the divisional officer who negotiated our first agreement either had not known about Mbinang's traditional role or didn't think it was significant.

In fact, the location of Sanaga-Yong Chimpanzee Rescue Center—I finally chose this name for the sanctuary because its location was near the confluence of the Sanaga and Yong Rivers—in the forest behind Bikol 1 gave this part of Mbinang's territory much more importance to the new chief. The sanctuary was considered a form of development, and the area would no longer be ignored. Soon after Chief Tendi Ibrahim was officially installed as the replacement of his uncle, he sent a letter to inform me that I was in his territory illegitimately. I was obliged to enlist the help of the divisional officer again to negotiate and sign a new agreement with the chief of Mbinang. This time we worked with the village people to draw up expanded boundaries, and Chief Ibrahim signed an agreement that prohibited logging, farming, and hunting within the 225 acres of Sanaga-Yong Chimpanzee Rescue Center. In exchange, I would hire all our staff from the villages, instead of bringing more qualified people from other parts of Cameroon, and continue to buy the food we needed for chimpanzees, staff, and volunteers from village farmers. I had already compensated the few farmers from Bikol who had been using plots inside our boundaries.

I abided by my part of the agreement and was a good neighbor in the community in other ways as well. I started providing medical care when I realized that people in the villages were dying for lack of simple treatments. The government hospital in Bélabo required payment in advance for consultations and all treatments, and there were no exceptions. People could die in the hospital parking lot for lack of a few francs. For people in our village community, I dispensed medications for malaria, respiratory infections, and diarrhea, and I allowed them to pay me afterward in fruit for the chimpanzees or by doing work around the sanctuary.

But some of the treatments I provided for illnesses and injuries weren't so simple. I sutured some terrible machete wounds, including one that was intentionally inflicted upon a man by his wife, and treated various other severe injuries, including third-degree burns in a child whose clothes caught fire. I often wished the people had much more specialized care, but I was much more skilled than anyone else they had, and I did my best. The care I gave saved a lot of lives. Ministering to human maladies took a lot of my time and distracted me from the work of the sanctuary. I sometimes resented it, but I kept my complaints to myself. The people who stood before me asking for medical help had no other options. I couldn't turn them away.

In return, I insisted that the people of the villages live up to their part of their agreement with me, but they were less than 100 percent compliant. One afternoon from camp I heard a chain saw cutting in the forest nearby, and thought it might be from inside our boundaries. I knew it wasn't Liboz or any of Coron's men, because they had moved on. I suspected the cutting was by illegal loggers we called "chain saw guys." These local chain saw operators followed the roads made by logging companies with legal concessions and cut smaller and/or more-difficult-to-access trees left behind by the big companies. Working in small groups of two or three, they could slice a tree into planks where it fell and carry the planks out of the forest on top of their heads to stack them on a waiting truck.

Kenneth drove me, armed with my camera, holding my head out the window to follow the growl of the chain saw. I strained to listen and gauge the distance and direction of the sound, and as it grew louder, I soon knew that the chain saw was indeed operating within our boundaries. Finally, when Kenneth and I were as close as we could get with the truck, we got out and continued through the forest on foot. The two men didn't notice our approach. One of them stood atop the large trunk of a downed tree, bending over as he worked the chain saw to cut the trunk in half. Just as he looked up at us, I snapped a photo. Quickly I turned the camera on the other man and snapped again—another good face shot. Whether or not they knew about my agreement with the village community, these chain saw guys had to know that it was illegal to log on government land without a formal concession. Even before the man turned off the chain saw and I could hear the words he shouted, I understood from disgruntled faces and pointing fingers that neither man was happy about the photos. When the noise died and we could hear their loud complaints, Kenneth didn't bother to translate, as he would have if he had questioned my ability to understand.

I held up my hand to ask for silence and confronted their anger with a quiet and polite question. "Can you please tell me for whom you are working?" Kenneth smiled casually as he translated.

Clearly surprised by this weird confrontation in the middle of the forest, they looked at each other, unsure whether to answer.

"Who sent you here?" I asked more directly, in simple French.

"The chief of Mbinang," the man with the chain saw finally responded.

Now I was the one who was surprised. As Kenneth and I had tracked the sound of the chain saw, I had played a scenario in my head where I would go immediately to Chief Ibrahim of Mbinang to enlist his help in putting an end to the illegal logging. To hear that he was responsible for it was unsettling. The chief had signed an agreement with me, formalized by a representative of the national government, promising not to log on our small part of the forest. I had enjoyed a glass of wine with him a few days earlier. I tried not to look as surprised as I felt.

"Okay, merci," I said, and to Kenneth, "Let's go!" I wasn't sure how strongly these men felt about the pictures in my camera. I didn't want to lose my beloved camera, but neither Kenneth nor I would have fought the guy with the chain saw over it. We hurried back to the truck before they had a chance to weigh their options and drove back to camp, where I could consider mine.

The next morning we drove to Mbinang. Chief Ibrahim sat on a bamboo bench in the meeting area, which was covered by a raffia roof in front of his house. With him were his nephew Alain, visiting from the town of Mbandjock where he worked as an administrator for a sugarcane plantation, and the two chain saw guys we had gone out of our way to meet the day before. After Kenneth and I had shaken everyone's hand, the chief gestured toward a vacant bench. While we beat around the bush with irrelevant conversation, as is customary and required before getting to any serious issues, other villagers wandered over. They shook our hands and then lined up around the periphery of the meeting area to listen in on what promised to be a confrontation.

Finally, I veered sharply toward the point. "Chief, we have a big problem!"

"Yes," the chief agreed.

"Yesterday, I found these men cutting a tree on sanctuary land! They said they were working for you!" I cut to the chase.

"Yes," Chief Ibrahim confirmed and proceeded to explain calmly. "I am not cutting wood to sell, as that would be illegal. I am taking trees only to make furniture in my house, which is my traditional right." He had been coached what to say, I felt sure, and I knew it was a lie. The chief's house was not in need of furniture. "But you signed a legal agreement, a promise, that you would not encroach on the sanctuary land for anything."

"It's my land," he said simply.

"Does your word, your *promise,* which you gave me in writing, mean anything at all to you?" I spoke in English with Kenneth translating, as I couldn't afford to be misunderstood. My words were direct, but my tone was respectful and calm.

Alain, who was a very large, relatively well-educated man to whom I had barely ever spoken, jumped into the conversation with a confrontational, angry tone. "The chief can do whatever he wants on his land. You have disrespected him and stolen his land!"

"The land is Cameroon government land, and the government sent me here." I was suddenly furious and my heart was pounding madly, but I struggled to maintain a calm tone.

"Look around you. Where are your government friends? You are here in the bush with us!" Alain spat at me. I had to admit, to myself at least, that he made a good point, but I was taken aback by how much this man I barely knew seemed to dislike me. When I didn't say anything, he pushed his point a bit too far, "The divisional officer and the national government in Yaoundé don't matter at all here."

"Well, I expect they'll be very interested to hear that you feel that way," I informed him.

"The chief is the only authority here!" Alain was animated and loud now.

"Well, if that's the case, it should be no problem for me, because your uncle, *the chief,* is the one who signed a legal agreement with me!"

"Taking photos of people without their permission is illegal in my country. You are the one who can go to jail." Apparently remembering he needed to address the grievances of the chain saw guys, Alain glanced at them, still scowling from their corner bench, as he spoke to me.

"They were trespassing. It's legal to take photographs of trespassers." I was making it up as I went along.

"It's the chief's land! You are the trespasser!"

"What do you know anyway? You don't even live here," I said, and I noticed that some of the villagers around us were nodding their heads a little as I was talking. Taking this as evidence that Alain wasn't universally liked in the village and empowered slightly by the subtle support, I went on. "I live here with the people and do many things to help this community. I'm part of this community. What do *you* do to help the people here? *Nothing,* that's what!" I really had no idea about his role in the community, but I was on a roll. "You live in *Mbandjock*!" I placed a derogatory emphasis on Mbandjock like it was the most ridiculous place in the world that a person could live, and people laughed, even before the translation.

Alain was so angry he couldn't stay seated. Standing added the emphasis of his large size to his shouting, both obviously intended to intimidate me. *"You are kicked out of the chief's territory. Leave now!"* He pointed to the road. *"Now!"*

"Who are you to kick me out?" I remained seated. As his voice had gotten louder, I made mine softer.

"I speak for the chief!" He still stood over me, and I tried to appear as relaxed as I could while I hoped he wouldn't try to physically throw me out in the road.

I looked at the chief. "Does he speak for you, Chief?"

The chief was silent, looking straight ahead, neither at Alain nor at me.

"Leave!" Alain commanded me, apparently interpreting the chief's silence as support for his position.

"Chief, does Alain speak for you?" I asked, a little more urgently.

Finally, the chief groaned uncomfortably and tilted the top of his head to one side and then the other. He obviously didn't like where this discussion had gone. "I'm adjourning this meeting." It was the first thing the chief had said in a few minutes. "Can you come back tomorrow at 10:00 A.M.?" he asked me. He had found a way to diffuse the situation.

After I agreed to come back, Kenneth and I quickly departed from the village, leaving without the customary good-bye handshakes. Back at Sanaga-Yong Center, I barely slept that night, worrying about the outcome of the meeting that would take place the next morning.

However, at 10:00 A.M. we arrived in Mbinang to a completely different atmosphere. As I approached the outdoor meeting place, Alain stood to shake my hand and addressed me solemnly as "Doctor." Before sitting, Kenneth and I shook the hands of Chief Ibrahim, the chain saw guys, and several onlookers who had gathered for the scheduled meeting.

Immediately after we sat down, Chief Ibrahim jumped to the point, "Can I finish cutting the one tree into planks?"

"Yes," I said, a bit reluctantly but in the spirit of compromise.

"Okay, then. I won't cut any other trees within your project boundaries." The chief settled it. I suddenly realized that he probably didn't intend to cut any more anyway, at least not in the immediate future, so it was an easy compromise for him, in principle only.

"Can we have the film from your camera with the pictures of these men?" Alain asked me with exaggerated politeness.

"It's a digital camera and doesn't use film," I explained, in an equally polite tone. "I will erase the photos."

"How will we know that you erased them?" Alain asked, staying respectful.

"You'll have to take my word for it, like I have to take the chief's word that he won't cut more trees on sanctuary land," I said. Alain and the chain saw guys were stuck with it.

There wasn't much more to say. The meeting was short. Before I could rise to leave, the chief asked if I would do him the favor of giving him a ride to a big meeting in the village of Mbargue (an hour away because the road was bad) the following day. It was hard to say no in that current context of reconciliation.

"Okay," I agreed, hiding my annoyance. I would send Kenneth to drive him, using our fuel.

Thus, the drama that had sucked my attention for two days ended anticlimactically. I had made the point that I would notice them logging on sanctuary land, but beyond that I hadn't gained anything. It was like many other experiences that defined my relationship with the people of the village; they were melodramas that demanded my attention and sometimes upset me, but generally ended okay. A few weeks later, a more serious incident in Bélabo traumatized me much more.

One night in September, I drove to Bélabo to pick up Estelle and two volunteers who would be arriving on the train from Yaoundé, leaving Kenneth and one other volunteer in camp. Kenneth and I might both have gone to Bélabo, but I didn't want to leave one volunteer alone. I might have sent Kenneth to pick up the two volunteers, but because I was excited that Estelle was coming for a short visit—her first in a long time—and because I hadn't been out of camp in a few weeks, I decided to drive to Bélabo myself, alone.

Two identical passenger trains operated each day on the railways running through our small town of Bélabo. One passed through on its twelve- to eighteen-hour route from Yaoundé to the town of Ngaoundéré in North Cameroon. The other followed the exact route in reverse, passing from Ngaoundéré to Yaoundé. The crowded trains, which traveled at night to avoid the heat, were both scheduled to arrive in Bélabo around 12:30 A.M., but one or both were often late. After checking at the train station and learning that the train coming from Yaoundé would be several hours late, I decided to rest in the small, sparse chamber of the Est Hotel that we rented for the equivalent of $16 per month. Renting a room by the month, we could keep our own clean sheets on the bed and our mosquito netting hanging above it.

Before going to the hotel, I visited a small kiosk near the train station to buy candles, matches, drinking water, some cookies, and a premixed bottle of gin and tonic. I figured I might as well relax and enjoy myself while I waited. When I pulled up to park across the street from the hotel, the attendant was standing outside the front door of the reception area under a dimly burning incandescent bulb. It was the only light source in the otherwise dark alley. Bélabo had electricity about half the time and rarely late at night, so I was surprised to see the bulb burning.

I climbed out of the truck with my hands full of keys, flashlight, and the plastic bag of items I had just bought at the store. I had taken no more than three steps toward the attendant when a hard blow on my back sent me sprawling painfully across the gravel. When I rolled over and tried to sit up, a man in shadow hit me across the top of the head with something he held in his fist, knocking me back down. I saw the flicker of stainless steel and realized he had hit me with the handle of a big knife.

He pulled the punch was my first thought. I realized immediately that he could have hit me harder or cut me with the knife, and it provided me with a reason to believe I might survive whatever was about to happen. Faceup, prostrate on the dirt, I opened my mouth to yell and was silenced by the sharp point of the long triangular knife blade on the base of my throat. The large blade rose up diagonally under my chin and connected to the hand of an African man who straddled me. In the darkness, I couldn't see his face or the faces of the other two men who stood on either side of me. The shiny metal of their knives was the only thing I could see clearly. A few yards away, under the aura of the dim bulb, I saw the hotel attendant go inside and close the door. I was on my own.

Without options to defend myself, at the mercy of these men whose faces I couldn't see, one thought ran through my mind like a mantra, *Stay alive, stay alive, stay alive, stay alive . . .* While their hands ripped off my money belt and my watch and searched my bra for money, their rough kicks bruised my arms and legs, but I didn't feel any pain. When they barked at me in rapid French, it sounded like gibberish through my fright. One of them yanked my blue Teva sandals from my feet, and another pulled my pants down to my knees. *They're stealing my good pants,* I thought. The terrifying idea that their intentions in removing my pants might be far worse came no sooner than it was laid to rest by one of the men searching my crotch for money. They were hateful two-bit thieves, and nothing more.

Satisfied that they had gotten everything of value, they turned away, leaving me shoeless, with my pants to my knees. One of them reached down to pick up my bottle of water, which had gone flying when they hit me from the back. Unhurried, he threw back his head to drink as he walked away. As I watched the three silhouettes saunter into the darkness, I tried to register as much information about them as I could. The one who held the knife on my throat was short and muscular, and he wore a red jacket. The two others—one tall, one of medium height—wore dark clothes. I knew I wouldn't be able to identify them.

As soon as they were out of sight, I stood up on the unforgiving gravel and began searching for my truck keys and my flashlight, trying to estimate where they would have landed. My heart raced as I looked about frantically, fighting an irrational fear that the bandits would come back. From within the sphere of dim light, I peered into the surrounding darkness and wondered if they were watching me. When I saw blood on the rocks, I realized that the broken glass of the gin and tonic bottle had sliced my foot, but I hardly felt it. After several minutes, during which I found my flashlight but couldn't find my keys, I banged on the hotel door and called to the attendant. I couldn't bear to be out there alone anymore. It took him several minutes to come, and I spotted my keys just as he tentatively opened the door. I had intended to ask him to go to the police with me, but I was suddenly suspicious of him. He had left me alone with the bandits. Was he simply afraid, unwilling to risk his own life for mine? I could understand that. He couldn't have even called the police during the attack, because Bélabo had no phone service. However, as he stood before me, it occurred to me that he might have cooperated with the bandits, and suddenly he seemed as sinister as they were. I fled in the truck as quickly as I could.

Alone in the pickup cabin, bumping along the forest road on my way back to camp to get Kenneth, intense feelings of relief vied with anger. My body was beginning to ache, but I had not been damaged in any lasting way. I was sure the thieves were quite disappointed with their booty. My money belt had contained the equivalent of $47, and my watch was an inexpensive Casio, the band of which they had broken when they yanked it from my wrist. I really needed my shoes and was upset about losing them, but the thieves wouldn't profit much—they probably couldn't sell them for as much as $20 on the streets. My biggest concern was the loss of my passport, which had been in my money belt.

I harshly berated myself for forgetting my many years of martial arts training. By the time I was on the ground with a knife at my throat, it was too late to respond. I should not have been caught by surprise. Seeing the hotel attendant rendered me complacent, and I didn't even look around before getting out of the truck. I screwed up, and I was lucky the price I paid wasn't much higher.

Kenneth and I drove back to Bélabo to pick up Estelle and volunteers Claudine Erlandson and Nicholas Bachand, and to file a police report about the bandit attack. A derailment, not a rare occurrence on our railways, had delayed the train by ten hours. It was on our way to the office of the military police that we spotted the blue passport book lying in the middle of the road several blocks from the hotel. The bandits had made a choice to discard it in the road, where someone might see it, where I might get it back. Although I was compelled to file the report, I was not optimistic about the bandits ever being captured. The military police also suspected that the hotel attendant was involved—if only due to coercion and fear. They would investigate.

When we returned to camp in the afternoon with Estelle and the new volunteers, nursery caregiver Ndele Chantal informed us that four-year-old Njode was sick. In fact, he was critically ill with pneumonia, and all my emotional energy for the next twenty-four hours was focused on saving him. Estelle and I passed the night in the nursery with him, one or the other of us watching his every breath. I was able to sleep for a few hours only because she was there. I trusted Estelle to detect even the slightest changes in Njode's condition, just as I would. She stayed another two days, until we knew he would survive.

I had no time to nurse my own psychological wounds, although I knew I had been affected by the bandit attack. I was more nervous than usual, constantly looking around behind myself, taking stock of my surroundings, visibly jumping when anyone walked up behind me. This fearfulness that came afterward was the worst part of the attack for me. For a few minutes the thieves had terrified me with their power over me, and now I was letting the memory of them terrorize me. It made me furious with them and with myself for not having more control.

A week later, I was in Yaoundé for meetings in the Ministry of the Environment and Forestry. This ministry was about to be divided into two, bringing the total number of government ministries in Cameroon to sixty-two. After the division, we would fall under the domain of the new Ministry of Forestry and Wildlife. It would take eight years and many meetings to propel our formal Protocol Agreement, the important document that would define the parameters of our working relationship with the government, through the labyrinthine bureaucracy of the ministry and finally receive the signature of the minister. I attended many meetings in the ministry offices, trying to move it forward.

Although cell-phone service wouldn't reach Bélabo for another two years, it had just arrived in Cameroon's cities and some of its towns. Using Estelle's new cell phone, I called my one friend from Cameroon on his. Since I had met him, George Muna had provided me with contacts, logistical assistance, financial support, and advice on cultural issues. Since we had moved the chimpanzees from his hotel, he and I had met several times. Whether by landline or cell phone I usually called to say hi and let him know when I was in Yaoundé, and most times he managed to make the trip to see me. He also liked to visit his brother who was living in the city, so he never lacked a comfortable place to stay, and this gave him a second reason for making the effort. Once he came to Sanaga-Yong Rescue Center to see what we had built for the chimpanzees. However, the camp accommodations to which I had grown so accustomed were more challenging for him. He seemed particularly averse to our pit latrine, and since he was older and not very athletic, I imagined he might lack the flexibility to meet its particular physical requirements. Why else would one hate it? A volunteer suggested that the flies and bees flying from it and the big spiders along its walls might be dissuading factors for some people. In any case, George only stayed in camp one night. Although I didn't see him often, I considered him a true friend, and it was his comfort I now sought.

When I told him on the phone about the bandit attack, his concern for me was palpable through the line. He insisted on driving the six hours from Limbe to Yaoundé to see me, even though I would be going back to Sanaga-Yong Center the next day. In George's arms, I felt safe to cry, and for the first time, I realized that I had feelings of love for this sweet man whose world was so different from mine.

thirteen

PREGNANCY AND MOTHERHOOD

Volunteer Claudine Erlandson, a French national who had married an American and lived in the United States for thirty years, was exactly my mother's age—born the same month and the same year. In October of 2000, she and I chatted as we drove to Bélabo for a shopping trip. The subject of children came up—she had two who were grown—and I confessed to her my twinge of regret about not having children.

"It's not too late for you, Sheri," she encouraged.

"Oh yes, it *is* too late," I assured her. "I made a different choice."

A few weeks later, at the age of forty-one, I discovered I was pregnant. George Muna, who was ten years older than I with two grown children and a teenage daughter from previous relationships, was the father of my baby. I hadn't been on birth control in Cameroon because I hadn't needed it in the forest. When I became involved with George, I thought it was very unlikely that I could get pregnant at my age. I wouldn't have made a conscious choice to do it, living as I was in the Mbargue Forest, with my irreversible commitments to the chimpanzees I had brought there, but on some level I really wanted a child.

Up until this point, I had successfully compartmentalized the different aspects of my personal life. Edmund was still raising all the money for the work we were doing in Cameroon, and I had recently convinced IDA to contribute a part-time salary for him to continue fund-raising. But with the communication challenges in Cameroon, he and I could go weeks without speaking. We had spent eight months of 1999 and all of 2000 living on different continents with little communication. We exchanged e-mails when possible, but those opportunities were limited. When I last had seen Edmund late in 1999, we didn't speak of fidelity. He was a gregarious, social person—much more so than I—and I assumed he was dating other people. I had made my choices and would live with them, but I still loved Edmund and cherished the idea of having him in my life, of knowing he worried about me, of knowing he would be there in the United States, at least on some level, when I went home. At the same time, I had come to rely on George's friendship in Cameroon and to love him too. On opposite sides of the planet, the two lives hadn't conflicted. I felt like a different person depending on which continent I inhabited. My pregnancy would bring my two worlds crashing together and change my relationships with Edmund and with George forever.

At Sanaga-Yong Chimpanzee Rescue Center, where I would spend the largest part of my pregnancy, I had to contend with Kenneth. "So you're the only one around here who gets to have a baby!" was Kenneth's response to the news of my pregnancy. It was the last thing he expected, and not only because of my age. When he had questioned my decision to use birth control in our chimpanzees, I had patiently explained my reasons with unwavering certainty. First, we needed to reserve our space and very limited resources for orphans in need of rescue. Second, I didn't think chimpanzees should be born to lives of captivity, no matter how special the form of captivity. Kenneth had been skeptical. With similar fortitude I had performed a spay (ovariohysterectomy) and a castration on our female and male camp cats over Kenneth's objections. To his thinking, there was a certain hypocrisy in my having a baby after I had denied that right to others.

While I was excitedly adjusting to the idea of motherhood and trying to work out how I could fit a child into my unconventional life, I was called back to the United States when my own mother was critically injured in a fall. The last couple of years I hadn't been able to speak with my mother very frequently, and I knew she had worried about me a lot. She didn't really understand enough about what I was doing and why to feel proud of me for it. For her, there was just the worry. Now it was my turn to worry during the two months I stayed with her in Mississippi, often sleeping in her hospital room.

During those weeks at the hospital, I was disturbed by my strolls through the maternity ward. While a baby chimpanzee, or even a baby kitten, could fill my heart with tenderness, these helpless and somewhat unattractive—to my fearful eyes—human infants left me completely cold. I began to seriously question whether I really wanted one. My doubt persisted until an ob-gyn at that hospital in Mississippi performed an ultrasound to confirm my pregnancy. Seeing my own tiny fetus's heart beating, like the fluttering wings of a fragile butterfly, settled the issue once and for all. Through an amniocentesis I learned that my little butterfly was a girl, whom I named Annarose long before she was born. With all my heart, I wanted her to thrive.

I didn't discuss my pregnancy with my mother. I knew that she wasn't overly conservative, and I knew she ultimately would be delighted about the baby, but I worried that she might experience some slight hesitation about my unwedded state, which I had no intention of changing. When my mother was conscious, I sometimes grappled with whether or not to tell her, and her strange hallucinations of babies in the hospital room with us was slightly unnerving.

"Sheri, put some clothes on that baby!" she ordered me on one occasion in her strong Southern accent.

"What baby, Mama?" I replied softly.

"That baby over there needs some clothes on it like the other baby has on!" she insisted, pointing to a dim corner of the room.

I turned on the light to convince her. "Look, Mama, there are no babies."

Another time, she told me to pick up the baby. When I again convinced her that the room we occupied was babyless, she laughed softly at herself. "I guess I need to get this baby stuff out of my mind."

I intended to tell her later, when she was stronger, that she would have a granddaughter. I thought the time would come for us to wonder and laugh about her "psychic" baby hallucinations in the hospital. However, her injuries and other health issues proved overwhelming. She died without knowing I would have a child. Losing my mother so early in her life—she was only sixty-two—was the deepest and saddest loss I had experienced in mine.

Immediately after her funeral, I flew from Mississippi to Oregon and stayed for three days in my house with Edmund, who was still living there and running our small IDA-Africa office out of it. For him my pregnancy marked a sort of sad milestone in our relationship, and the strain of it was between us. Nevertheless, he did his best to be supportive and comforting to me in my grief. He tried to make sure I ate well, at least, during those few days before I took my grief back with me to Cameroon.

Back at Sanaga-Yong Center, another source of sadness for me was my estrangement from Pepe. His problem with me started just after I returned from the United States, and it continued throughout my pregnancy. Whenever he saw me, he displayed aggressively toward me, so that I couldn't even approach his cage. If I was with other people, he met my eyes during his angry display, making sure I knew his problem was with me specifically. With my deeply hurt feelings, I wondered whether his hostility had to do with me leaving for so long when my mother was injured, or whether he could smell the hormones of my pregnancy. It started before he could *see* any signs of my pregnancy.

George visited me in camp from time to time. His visits were brief but I was always happy to see him. Initially he was excited about the baby, then he grew concerned about whether I could be a good mother with all my other responsibilities, and eventually, he was excited again. When he visited, he brought me pastries and pizza and olives, and we talked about issues concerning my staff or the villagers that confused me. Our discussions were enlightening and his advice always useful. Once he told me, "Sheri, you're too direct for our culture. You need to beat around the bush more." He loved to use American clichés—he had gone to university in the United States for six years. Once George traveled with me to Yaoundé for an ultrasound. I suffered from motion sickness on the train, which I had never done in my nonpregnant state. The bathroom was filthy, and George stood over me and held my hair while I vomited over and over on the rocking train. The quality of the ultrasound performed in Yaoundé wasn't good, but we could see Annarose's face and her tiny fingers. It was thrilling to share that with George. However, to a large extent, his business kept him occupied in Limbe, fourteen hours away from me, and there was no phone service anywhere around Sanaga-Yong Center. I had brought back a satellite phone from the United States, but it was much too expensive to use casually. Mostly, I was on my own.

All in all, the months of my pregnancy proceeded pretty much the way the months before it had. Problems and crises showed no deference to my maternity. Late one afternoon about six and a half months into my pregnancy, I realized that Becky was seriously sick. It was already four o'clock when caregiver Assou informed me that Becky hadn't eaten since morning and had returned to her cage from the forest. I found her resting on her tire, which was situated on a platform high above my head in the center of the cage. A year and a half earlier, I had positioned the tire on the platform and secured it there with chain, thinking it would make a comfortable nesting spot for someone. Becky had chosen it as hers, exclusively. She often carried in leaves and vines from the forest to make a nest on it, but today, she rested on her side on the bare tire. Her back was toward me so I couldn't see her face, but I could just barely discern the much too rapid rise and fall of her abdomen as she breathed. I called to her for over ten minutes, and finally, as I was failing in the fight against my escalating sense of panic, she descended and came to me. Normally, her intelligent, questioning brown eyes would have found mine during her approach, but today, under deeply furrowed brows, they looked past me. I knew immediately that she was in terrible pain.

"What's wrong, Beck?" I reached in to place my hand on her forehead. Her skin was warm. When my hand moved downward to briefly cradle the side of her face and then skim gently over the front of her chest, she seemed to hardly notice, but when I touched her abdomen, she grunted and pushed my hand away. The problem was in her belly. She turned her back to me and allowed me to stroke and groom her back for a few moments, but it was more comfort to me than to her. As she left me and climbed slowly back up to her tire, I noticed a slow dribble of urine that continued after she was lying on her side again. Deeply dreading the prospect of performing surgery on Becky, I began considering it. Whatever was obstructing Becky's urine and causing her such awful pain was not likely to repair itself or to be fixed by medicine.

I used the satellite phone to call veterinarian Jim Mahoney in the United States. After working with chimpanzees and monkeys in biomedical laboratories for years, Jim had been the driving force for placing many of them in North American sanctuaries a few years earlier. I knew he had gained a lot of medical experience that

I lacked, and I appreciated that he had always been generous with his advice. Jim answered his phone and listened attentively while I explained Becky's symptoms. On the basis of what I told him, he suspected that Becky had twisted her large intestine, and that the twisted bowel was blocking the flow of her urine. He said he had seen it before in chimpanzees. If he was right, the blood supply to Becky's intestine was being cut off, and she probably couldn't survive until morning without surgery. Jim warned me that I might have a fatal outcome even with surgery. If he had known the difficult conditions under which I would operate, I suspect his pessimism would have been much greater.

I had received funding for a small veterinary clinic from the Michigan-based Arcus Foundation, but I had not built it yet. Our kitchen table would serve as my makeshift surgery table, where I had already performed a few minor surgeries like wound suturing. It was a round slab of wood, sliced from the end of a big log and situated under the raffia roof we called our kitchen. I had no gas anesthesia or skilled technicians to assist me. Flashlights charged on our solar system would be my surgery lights.

By the time I had sterilized my surgical instruments in our pressure cooker and converted the open-air kitchen area to a surgery suite, as much as possible, it was dark. Getting Becky out of her cage would be the first challenge. She was a big girl, weighing about 130 pounds, and I was hoping she would descend from her tire-bed to the ground when I blew the dart of anesthesia into her thigh. Unfortunately, she was so sick she hardly moved. Kenneth climbed a ladder up to her platform and somehow managed, from his precarious perch on the ladder, to pull Becky off the tire and hand her down to the night watchman and me. We were careful to stay out of reach of Pepe and Jacky, who barked and screamed at us from the adjacent cage chamber where I had locked them. Not understanding how we could, or why we would, be taking Becky from the cage, they were frightened for her and wanted to protect her. Hanging on the cage wall, Pepe applied the full force of his fury to loudly vibrate the metal mesh in a threatening display that succeeded in terrifying the night watchman, and in scaring Kenneth and me as well. I knew we had built the cage well—I was there when all the pieces were welded together—but I breathed a sigh of relief when we managed to get outside the cage with Becky and lock the door behind us, putting another wall of metal between Pepe and us. We placed Becky on a bamboo bench that served as our stretcher, and through the swath of light radiating from my head torch, we carried her along the dark forest trail to the surgery table.

She was too big to fit on the log table, so Claudine sat on a stool at one end, cradling her head. Using a vein in Becky's back leg, I started an intravenous fluid line to improve her blood pressure and start potent antibiotics. Through the IV line, I would administer anesthesia injections as necessary throughout the surgery. Nicholas Bachand, a Canadian volunteer who years later would become a veterinarian himself, and Kenneth assisted me during the operation—holding the flashlight on my surgery field and handing me items as I needed them.

When I cut into Becky's abdomen, I found that all the organs in her abdomen were glued together by fibrous adhesions; I had never seen anything like it before. I couldn't even identify the individual organs until I broke down the adhesions, which was a painstaking process that took hours. One wrong move and I could have caused bleeding that would have been impossible to control in the large mass of adhesions.

With my advancing pregnancy, I had become prone to low back pain. Leaning over the low log table for hours on end, my back now hurt so badly I wanted to scream. Instead, I cursed and swatted at flying insects attracted by the flashlight held over Becky's abdomen. The tone of my instructions to Nick and Kenneth reflected my agonizing discomfort and stress.

"Tilt the flashlight more this way!"

"Shine it where my hands are working! I need to *see,* dammit!"

"Wipe the sweat off my face before it drips into Becky!"

"Give that syringe of ketamine! Quick!" (Just before starting the surgery, I had taught them to give an injection through the IV fluid line.)

Several times, I had to put down my surgical instruments to draw up doses of anesthesia or change to a new bag of intravenous fluids, leaving Becky's blood on everything I touched. Each time, I tossed off contaminated gloves and put on a new pair before continuing the surgery. When I had finally separated out the intestines and other organs, I could see that Jim was exactly right. Becky's large intestine was twisted and beginning to turn purple. After I untwisted it and tacked it down in its normal place so that blood circulation could improve, the pink color began to return, giving me more hope that Becky might survive if we could get her through the surgery.

That Becky lived to the end of that long and complicated surgery under such ridiculously primitive circumstances was a testament to her amazing strength. To survive beyond it and recover her health would be another battle. It was dawn when we finally moved Becky to a chamber in the satellite cage that Dorothy and Nama had lived in during their early months at the center. I would need to keep her separated from the other chimpanzees until her surgical wound healed. Kenneth left me in the cage with Becky and returned shortly with the hot water bottles I had requested. As I strategically placed the hot water bottles and tucked the blanket around Becky's cold body, I began to cry. Sitting with her in the cage, praying for her to wake up, feeling I had done all I could, I sobbed with abandon. I had never been too much of a crier, but pregnancy changed that for a while.

"Ma'am, she may survive," Kenneth said to comfort me.

"I knooow," I wailed at him.

Later that morning, with Becky awake and stable, I thanked Kenneth and the exhausted volunteers for their stamina. I also apologized for the nastiness they had endured from me during the surgery and the hours leading up to it.

"Don't worry, dear Sheri," Claudine assured me in her sweet French accent. "You didn't yell at us *too* much." I sincerely hoped I hadn't.

Three weeks passed before I was confident that Becky would survive the ordeal. Jim was always available to take my calls and consult with me during Becky's recovery, as he had been the night of the surgery. He speculated that migrating parasites could have caused the adhesions in her abdomen, and that the adhesions predisposed her to the twisted intestine. I knew I had to get antibiotics into Becky at all costs, and twice, when she refused to take them orally, I darted her with them, knowing that it would hurt her feelings and infuriate her. Fortunately, she wasn't prone to grudges against me.

One afternoon, while I was still engaged in the life-or-death struggle to get antibiotics into Becky, the young woman I had hired to tend our garden informed me that Cathy, our cook, needed me down in the small guard shed by the entrance to the camp. I found Cathy lying on the dirt floor of the shed in her traditional green-and-orange African dress, about to give birth to her fourth child, an event she had expected to occur a month later. I could see that her water had already broken. I had helped mothers of various species deliver their broods—kittens, puppies, calves, and once a foal—but never a human. When Cathy saw my distress, she told me she wanted to go to her village, and I certainly wanted to deliver her into the more experienced hands of her mother and the other women of the village as soon as possible.

While I collected a thin foam mattress from my cabin and spread it out in the back of the pickup truck, I sent the gardener to collect Stephanie DuSauccy, a pretty blond French volunteer in her early twenties, who would drive us to Cathy's village of Meyene. Cathy was able to climb into the back of the truck with minimal assistance, and I followed her in. As we bumped along the dirt road, Cathy lay back on the mattress biting her lower lip, and I, sitting between her legs, could see that the baby's head had crowned.

"Ne pas puissez, Cathy!" I told her, stupidly. *Don't push!* It was an impossible request.

"Oh, Madame!" That exclamation of warning was the only sound Cathy made as she pulled her dress up to her chest and expelled her baby girl into my hands. The tiny gray baby began to cry almost immediately, which I knew was a good sign. As we continued to bump along, I struggled to hang on to the slippery, crying baby who was still attached to the placenta, which was still inside Cathy. I hadn't thought to bring a towel.

When we pulled into the village of Bikol a few minutes after the baby was born, several women were waiting for us in the road. None of the men of Bikol could be seen. The women must have received the news that Cathy was in labor and, hearing our approach, had assumed Cathy was in the truck. I frequently marveled at how news could travel so quickly in these villages without phone lines. It was only when Stephanie hopped out of the driver's seat that she realized the baby had been born. She joined five women who surrounded the truck while two, including Cressance, the wife of Chief Antoine, crawled into the back with Cathy and me. After they gave Cathy a drink of water, they helped her to a squatting position so she could push out the placenta. While she pushed, one of the women outside the truck offered encouragement, or distraction, by tapping on the top of her head with a soft broom. I didn't really understand the broom's role, but I trusted that they knew what Cathy needed much better than I did. Soon I was holding a slippery placenta along with the slippery infant. When I asked for a string to tie the umbilical cord, Cressance smiled and told me to be patient. All the women, including Cathy, were more relaxed than I was.

Cathy said we should continue to Meyene, and Stephanie climbed back into the driver's seat to comply. I knew the wind was cold on the still wet baby, so I hugged her to my chest, thoroughly soaking my thin cotton shirt with blood, while I tried to keep the placenta, about as big as the baby, close to her little belly. As we approached the outskirts of Meyene, we saw Cathy's mother walking hurriedly on the road toward us. She was wearing a clean dress, instead of her farming clothes, and she wasn't carrying a machete or a basin as she would have been if she had been heading to her farm. I knew that she was on her way to Sanaga-Yong Center to help Cathy. We stopped so she could climb in the back of the truck with us, and seeing that both her daughter and new granddaughter were alive transformed her worried expression into a broad grin. She didn't sit down in the truck, but instead broke into a joyful song; she accompanied her own melodic Bam-véle lyrics with a slow twirling dance, reaching for the sky in gratitude. Leaning against the corner of the truck bed, just behind the empty passenger seat, Cathy smiled contentedly as she looked up lovingly at her mother. In that moment, all was right with the world of women in the pickup truck, and I laughed out loud to be part of it.

When we pulled into the village at last, a middle-aged woman I hadn't seen before reached over the side of the truck to take the baby from me. With the baby in one hand and the placenta in the other, she lifted both over her head as she turned away from me to walk toward Cathy's mother's house, where I hoped they would have a clean tool for cutting the umbilical cord. Cathy's mother and Cressance helped her out of the truck.

"Merci, Madame," she said, looking back at me.

"De rien," I replied just before climbing into the passenger seat of the pickup. *It's nothing.* It was the customary response.

Dorothy (in profile) was always kind, patient, and forgiving.

▲ As personnel manager, Raymond Tchimisso was called "Le Grand Chef" ("the Big Chief") by sanctuary staff.

▶ *Top:* Agnes sits beside Marylise and the baby named after Agnes.

▶ *Bottom:* The staff of the sanctuary. Sheri delivered the baby on the far right. Annarose is being held by her babysitter.

▶ Sheri, who is pregnant, cares for Gabby, who is sick.

▼ Sheri, six months pregnant, carries Caroline.

◀ Sheri, with Annarose, collecting wood chips to use as bedding for the chimps.

▼ The one-room cabin at Sanaga-Yong Rescue Center that Sheri and Annarose shared.

Annarose, George, and Sheri, during a brief visit to the camp by George.

Dorothy with the chimp named Bikol, whose name means "the King" in the Bamvéle dialect.

Top: Hope tries to stop Simon and Gabby from fighting.

Bottom: Emma, Njabeya, Mado, and Future, grooming.

▲ Dorothy, giving Gabby a ride with Bouboule on her left.

▶ *Top:* Grooming as a pleasurable community event. Manni watches Bouboule grooming Jacky.

▶ *Middle:* Sheri, Tilly, and Shy, who is eating a papaya.

▶ *Bottom:* Nama and Emma take pleasure lounging in the dirt.

Once Dorothy had adopted him, Bouboule stayed close to his new mother.

Now that he had a mother to nurture and protect him, Bouboule could relax.

Like the chimps did with one another, Sheri established and maintained friendships through grooming.

◄ *Top:* Sheri and a grown-up Gabby speak to each other through the fence.

◄ *Bottom:* Over time, Becky became gentler, and her relationships with both humans and other chimpanzees grew more harmonious.

▲ Dorothy, whose funeral would one day teach the world about the depth of animal feeling.

Dorothy and Bouboule, even when he was an adult, maintained a close relationship.

The people of the village came to the camp en masse to pay their respects to Dorothy.

Dorothy's funeral. Sheri cradled her head while the other chimpanzees gathered to view her body.

Jacky became a definitive leader and helped keep the peace among the chimpanzees.

Women in the village didn't pamper themselves during pregnancies, and it was not rare for childbirth to catch them by surprise while they were farming their fields or carrying produce to the market. Although Cathy's style of delivery wasn't out of the ordinary in the village community, being part of it was bizarre for Stephanie and me. I looked into Stephanie's blue eyes, and she stared back into mine. That we were mutually stunned was vividly conveyed by our silence and slack-jawed expressions.

She broke the silence. "We better get back to camp," she said, gesturing to my blood soaked clothes.

I simply nodded. There was really nothing much for Stephanie and me to say, but neither of us would ever forget the experience.

A month later, I left Sanaga-Yong Chimpanzee Rescue Center to give birth to my daughter in the United States. I knew I wouldn't be as brave as Cathy had been, and I wasn't. Whereas Cathy's labor had come earlier than expected and lasted less than an hour, my painful ordeal came more than two weeks after my "due date" and lasted twenty-four and a half hours. George accompanied me as far as London, but because his father was critically ill, he wasn't able to come all the way to the United States for the birth. Instead, Edmund was with me when Annarose was born, and he loved her from the beginning of her life. I truly loved him during that time, too. I didn't know what I would have done without his friendship and support.

Armed with several books on how to care for infants, I brought Annarose to Cameroon when she was seven weeks old. After a few days introducing George to his daughter in Yaoundé, I returned with her to the Mbargue Forest. I was completely in love with my beautiful baby daughter and vulnerable in a way I never thought possible. She was the one person I knew I couldn't bear to lose. I worried about bringing her to the forest, fearful that she might get sick, but I had committed myself to the chimpanzees before I even dreamed of her. People in the village raised children with far fewer resources and options than I would have. I belonged in the Mbargue Forest, and Annarose belonged with me. In my tiny one-room wooden cabin I built a baby-size extension on the side of my bed and slept with her close to me under my mosquito net. I learned later about the controversy in the United States and other developed countries surrounding mothers sleeping with babies, but I didn't know anything about that then. For me, there in the Mbargue Forest, it was intuitive—the way primates naturally mothered.

I allowed all the chimpanzees to see Annarose without getting close enough for them to touch her. The adults, even Pepe, were calm and quietly interested, while some of the juveniles, especially Bikol and Gabby, greeted us with little stomping displays. I thought they were jealous, and I was very concerned they might hurt Annarose. To keep her safe from injury and to prevent the exchange of diseases, I mostly kept her physically separated from all the chimpanzees.

I hired a young woman named Helene from the village to babysit in the camp while I worked throughout the acreage of the sanctuary. Several more years would pass before I equipped our staff and myself with walkie-talkies. When baby Annarose cried, Helene sent another employee—usually the cook or the groundskeeper—running to collect me.

"Madame, elle a faim!" *She's hungry!*

Many times a day, whenever Annarose needed to eat, I responded to the summons to breast-feed, and like the women of the villages, I never associated the slightest need for modesty with the function. It was as natural as me eating. Although we were in the forest with little electricity and no running water, I thought my experience of motherhood was easier, and possibly more rewarding, than that of many working mothers in the developed world—especially in the United States, where the three-month maternity leave was standard. I sympathized with mothers who had no choice but to leave their young infants to return to work. I was able to move forward with my work for the chimpanzees while keeping Annarose only a short jog away.

Earlier in the year, during a two-week trip back to his village in Northwest Cameroon, Kenneth had married his childhood sweetheart in a traditional ceremony. Within a month of my return to Cameroon with Annarose, he left the Mbargue Forest to begin a life with his new wife. It came as no surprise to me. He had wanted to leave soon after he got married, but he had agreed to wait for me to have my baby and get back to Cameroon. It had been difficult to imagine life at the center without Kenneth, although I had been trying to adjust to the idea of managing with volunteers and my local staff. I certainly had no choice but to make it work.

While I was grateful that Kenneth had stayed as long as he had, and I understood completely why he was leaving, I experienced an illogical sense of abandonment when it was time for him to go. He was going to a new, more normal life and leaving me alone to manage this strange one we had created together in the forest. The sanctuary had been my vision and he my employee, but I couldn't have started it without him. And he was the only one who would ever understand the effort and sacrifice of those first two years in the Mbargue Forest. During our final conversation outside the train station, Kenneth tried to reassure me: "You'll be okay, ma'am."

"Of course, I'll be okay," I told him in a forced casual tone, which I could see didn't fool him. Finally, I didn't try to hide my sadness. I avoided his eyes as I managed to say in a hoarse voice: "I definitely will be okay, but your leaving marks the end of something that was special."

"It marks the end of the beginning of your chimpanzee sanctuary," he said. Now I met his eyes as we both smiled at his well-timed truism, and it was a good time to hug good-bye.

fourteen

DOROTHY FINDS HER STRENGTH

I n January 2002, I decided to introduce a first group of juvenile chimpanzees—ranging in age from two and a half to four years—into Jacky's group of adults. Bikol, Gabby, Bouboule, Moabi, Njode, and Mado composed the group of six. Mado was the only girl. We always referred to this group of first juveniles at the sanctuary as "the babies," but they were really more like very strong and agile toddlers, weighing from twenty to forty pounds.

Free-living chimpanzees have been known to kill babies from different groups, and sometimes even from their own group, as have captive chimpanzees, but I knew that some adult chimpanzees at other sanctuaries had welcomed young orphans. We had introduced six-year-old Caroline, the chimpanzee who road in the cab of the truck with Estelle and Kenneth when we brought Dorothy and Nama from Luna Park, to the adults a year earlier. At one time Caroline had lived in the nursery with the other juveniles, but being a few years older, she had seemed bored with them and had become difficult to manage in the group. During our long daily excursions through the forest with the juveniles, Caroline sometimes opted to amuse herself by wreaking havoc in our camp instead. When we finally introduced her to the adult chimpanzees, they accepted her easily, so I was hopeful that our efforts to integrate the others would go well too.

One morning, while Jacky and the other adult chimpanzees were in the forest, caregivers Ndele Chantal and Mvoku Samuel along with volunteers Karen Bachelder, Mirjam Schot, and Gabriela Shuster carried the six babies from the nursery, walking single file along the forest trail, with me following closely after them with my video camera. When we reached the big chimpanzees' satellite cage, the loving porters entered a single chamber of the cage with the babies they carried.

With the exception of Mado, who had been surrendered to us more recently by a virology laboratory, the babies had been living together at Sanaga-Yong Chimpanzee Rescue Center for almost two years and had developed strong bonds between them. They had been coddled and protected by caregivers and volunteers, who took them on long walks in the forest every day. Now I hoped that the juveniles would support one another through this integration process and that the adult chimpanzees soon would become their new role models and protectors.

Dorothy, who usually sat at the edge of the forest instead of going deep inside it like the other chimpanzees in the group, was the first to become aware of the commotion at the cage. Although the babies had been hearing the adults vocalize, they had not actually seen an adult chimpanzee since poachers took them from the forest. When Dorothy approached the cage from the forested enclosure, curious but not agitated in the slightest, all the babies climbed down from human arms to move a little closer to the cage wall and get a better look at the big chimpanzee on the other side of it. During this moment that the babies had turned their attention elsewhere, all the human carriers slipped out of the cage, and we closed and locked the door behind them. I worried that the babies would have tantrums when they realized their human surrogates were gone, but no tantrums were thrown. The babies' attention was elsewhere. Within moments, Jacky, Nama, Pepe, Caroline, and Becky, who now seemed in perfect health, were on the scene, reaching their arms through the sizable holes in the metal mesh of the cage wall, trying to touch the frightened babies, who huddled, grimacing and fear barking, just out of reach in the chamber.

They were relatively safe, isolated as they were, in the cage chamber, but the holes in the mesh of this older cage were much larger than those in the improved version I would build for future cages. Even the large adult chimpanzees could push their big arms in as far as their biceps. Grabbing a juvenile's hand or foot, they could have pulled his or her whole arm or leg through the mesh. I was mostly worried about Pepe and Becky, whose reactions to newcomers were less predictable, and in those first hours, it was Pepe and Becky who solicited interaction with the babies most eagerly. To my great relief, while intensely interested, they were also calm and patient, without a single aggressive display or bark. To a large degree Pepe and Becky hogged the area along the front of the cage, while Jacky and Nama split up to watch their interactions with the babies from either side. Dorothy found a comfortable observation spot on a rise in the dirt a few yards away, and Caroline receded to watch the action from a tree limb at the edge of the forest.

The babies were fascinated enough that, within the first hour, four-year-old Bikol was breaking the huddle to make tentative movements toward Pepe and Becky, and soon the others were taking turns doing the same. It was a noisy affair. The babies obviously wanted contact with the adults, but when contact actually occurred, they screamed, slapped at the big arms, and ran back to the safety of the huddle. The nervous humans who had nurtured and loved the babies all stood by a few yards from the cage, ready to give moral support if ever it was solicited. Once, smallest baby

Gabby broke away to extend his arms toward me through the cage, asking for a hug and reassurance. "It's okay, little man," I told him as I passed my arms through the cage mesh to embrace him tightly. Ten seconds in my arms was all he needed before scooting back to his friends. Mostly the attention of the babies rested on the adult chimpanzees, and not once did anyone reach for us crying as though he or she wanted to be taken out of the cage.

Finally, during one of Bikol's halting approaches to Pepe, he turned suddenly to back up all the way to him submissively. Seeing Bikol's tiny frame juxtaposed with Pepe's comparatively massive one, I realized again how vulnerable he and the others were. Pepe hugged him gently from the back, while brave little Bikol turned his sweet face toward us for reassurance. It was a breakthrough. Tentative overtures continued through the afternoon, but with slightly less noisy fanfare.

For three weeks, I kept the six recently transported juveniles in this single chamber of the cage, adjacent to the two other chambers where the big chimps slept. Before we would let any of the adults into the babies' chamber, I wanted to give them enough time to get to know one another through the cage wall. But how much time was enough? I had almost no experience integrating adult and juvenile chimpanzees. I would watch them as perceptively as I could, read their reactions, and let them tell me how to move forward. Even so, I knew there would always be a degree of unpredictability. If anything, I would err on the side of being too cautious.

My maternal instincts were in full swing, and back in camp my baby daughter was thriving. When I was busy working, which was usually the case, she was riding contentedly on her nanny Helene's back, tied there securely by a square of colorful cloth called a pagne, which I had bought in Bélabo with the intention of using myself. Helene taught me to wrap the cloth all the way around Annarose and myself and to tie the four corners tightly, up high on my chest. All the women of the village kept their hands free for work by tying their babies on their backs in this manner. Unfortunately, I couldn't breathe well with the pagne tied around my chest, and I was always nervous that Annarose would fall out of it. I lamented half seriously that she wasn't learning to cling to me as a baby chimpanzee her age would be doing, as the baby chimpanzees in our nursery did when I took them to the forest. Apparently, the clinging or grasping ability didn't offer a strong survival advantage to the babies of bipedal hominids with free arms to carry them (or tie them with pagnes), and therefore the evolutionary process didn't select for it. I thought it was a regrettable loss. But for Helene and Annarose the pagne seemed a natural fit. Annarose often slept with her beautiful little head against Helene's back, where I imagined she was soothed by the sound of a beating heart. She could already hold her balance sitting up, and she had a joyful belly laugh, which George elicited for the first time making goofy faces during one of his brief visits. I was stunned to hear her laugh out loud for him, her shining brown eyes glued to his face, when she had not yet done so with me. It was my first inkling that Annarose would have an adoring relationship with her father that was quite independent of any involvement with me. She was in direct contact with an adult human almost twenty-four hours a day. It made me sorry for children around the world who had less care and even sorrier than I was before for our young chimpanzee orphans. I wished that each of them had a mother who would, if necessary, die trying to protect them. But for them, in this terribly unnatural sanctuary circumstance, there was only me, trying to make decisions that would secure a better future for them without endangering them too much.

Because Jacky and Nama had emerged as leaders of the small group, I wanted to encourage a strong bond between them and the babies. After a few days, I let the two of them enter the juveniles' chamber first, and over several more days I was delighted to observe their sweet playfulness and intuitive gentleness. Even while Jacky and Nama were inside the chamber with them, some of the babies chose to play with Pepe, Becky, or Caroline through the mesh wall. Jacky and Nama monitored the interactions carefully. Caroline was playing with her old friends like we would have expected from another kid; integrating them back with her would be no problem. Dorothy groomed all the juveniles sweetly, Becky kept inviting the eager little males to copulate with her through the cage mesh, which they happily did even though all were far from puberty, and Pepe loved to make them all laugh hysterically with his tender tickling.

One afternoon I was watching when Pepe became too persistent in tickling the back of Gabby's thigh, a place where many chimpanzees are ticklish. At two and a half years of age, Gabby was our youngest baby, and I thought he was the most vulnerable. As Pepe held on to his small ankle and pinched playfully at his thigh, Gabby's reaction turned quickly from squirming laughter to a frantic struggle to get free. When he screamed, Jacky rose immediately to intervene, and Nama too appeared ready to pounce. However, before either of them had time to act, Pepe willingly responded to Gabby's scream, releasing his leg and looking tenderly at his backside as he scooted away. After a few moments' respite out of Pepe's reach, Gabby returned to be tickled again. Pepe had showed me that he meant no harm, and in any case, Jacky and Nama were willing protectors. I decided then to open the doors and let them all go together into the forested enclosure the following day.

When we opened the sliding doors, the big adults all welcomed the vulnerable and trusting juveniles with open arms. After a half hour of hugging and playing just outside the cage, the kids seemed willing to follow these adults they admired anywhere. Soon, the whole group moved together along a trail leading into the forest, out of my and the caregivers' sight. I was happy to see that even Dorothy went along. Three weeks had passed since we moved the babies from the nursery, and the integration had gone perfectly so far.

A month later, it seemed as though the group of twelve had always been together. Nama and Jacky rarely carried any of the babies, but they played with them, and more important, they comforted, protected, and mediated conflicts without discrimination. The caregivers called Nama "la mère du monde," *mother of the world*. She and Jacky broke up skirmishes between the juveniles and meted out discipline justly—always seeming to know who needed the scolding. Pepe and Becky both let young Gabby ride on their backs, and he was the only one who ever had the privilege of sharing Becky's tire-bed. Dorothy was always gentle with the babies, but their roughhousing seemed to get on her nerves. She enjoyed grooming them when they were calm, and she tried her best to ignore them when they were rambunctious.

However, within another month I noticed that four-year-old Bouboule, who had been a clingy baby with his human surrogates, was now always with Dorothy, and she was grooming and kissing him frequently. When they sat together, rarely could any light be seen between them, and when she rose to leave, she often paused so he could jump on her back to ride. When she didn't want him to ride, he walked beside her with his arm draped across her back. Only when Dorothy was resting comfortably did Bouboule venture from her side to play with the other juveniles. He had found the mother he needed so badly.

Dorothy not only coddled and nurtured Bouboule, she exerted herself to protect him. If a conflict with another juvenile wasn't going his way, he ran to Dorothy's arms or hid behind her bulk. A soft bark from Dorothy would quickly dissuade any young pursuers.

Becky didn't seem to like Bouboule much, and once, irritated by his rowdy behavior near her, she barked and hit at him. It was a mild disciplinary rebuke, but Dorothy didn't like it. She spontaneously unfurled a back fist punch that landed squarely on Becky's upper arm. I winced with worry, waiting for Becky's wrath, surprised that Dorothy had invited it. Indignant and angry, Becky turned to scream at Dorothy, but the latter wasn't cowering in fear as she had done so many times before. She was screaming right back at Becky, their faces inches apart. Jacky and Nama stayed out of it, and the two female chimpanzees soon quieted down, without any biting or any blows thrown except Dorothy's first one.

The social dynamic had changed. In her forties, Dorothy's role as a first-time mother had transformed her. Whereas she had not been emotionally capable of defending herself against Becky, she could and would assert herself to protect her adopted son, Bouboule. Her newfound assertiveness elevated her status in the group, earning her a universal respect she had not enjoyed before. She and Becky would eventually become the best of friends.

fifteen

THE UNSPEAKABLE

In late September 2002, I was in the town of Bélabo collecting our weekly supplies of soap and bleach and some food items, such as rice and pasta, that we couldn't buy from the villages when volunteer Agnes Souchal used the satellite phone to call me from camp.

Agnes was a French national who had arrived to volunteer at Sanaga-Yong Chimpanzee Rescue Center a few months earlier. She was thirty years old, although she looked younger, and no more than five foot two. She had stunning hazel eyes and wore her dark brown hair in a very short style that wouldn't have flattered most women but was a nice frame for her classically pretty face.

Agnes's approach to the work was serious and cautious, as she wanted to do everything correctly. She asked a lot of "what if" questions, expecting me to have a plan for every possible contingency. This unnerved me a little bit, because I dealt with issues as they came up. I kept my eye on what I wanted to accomplish and visualized it happening that way. Inspired by my love for the chimpanzees, I had had the single-minded determination to push my vision through, but I had made some mistakes along the way. I was beginning to realize that more caution and anticipation of pitfalls could serve me well. It had been a year since Kenneth had left and I needed a manager I could trust to replace him. I wondered if Agnes's personality might be a good complement to mine. She was unusually smart, and I sensed in her a unique strength of character combined with gentleness. I really liked her.

A month earlier, Agnes couldn't have contacted me by phone in Bélabo because mobile phone service had just arrived, but she reached me that September afternoon with bad news. Pepe had fallen from a tall tree. The caregivers had seen him fall an hour earlier and hadn't seen him since.

Without regard for the suspension system of our old red pickup, I flew across the bumpy dirt road leading from town to camp, making the trip that usually took over an hour in under forty-five minutes. Pulling into camp, I stepped down from the truck with my heart pounding even before I ran as fast as I could down the trail through the forest to the enclosure. I found Agnes and caregivers Akono and Assou waiting for me outside the enclosure, as close as they could get to where they had seen Pepe fall.

Knitting her brow with worry, Agnes pointed up high to the limb of the tree just inside the forest from which the caregivers had seen Pepe drop. I bent my neck back to see the limb near the top of the tall tree. "My God! It must be a hundred meters." I said, anguish bleeding through my words.

Akono and Assou had both seen what happened.

"Out on the limb, Pepe swung a stick to hit Jacky. When Jacky grabbed the stick, Pepe lost his balance," Assou explained. "Then Jacky must have let go, because Pepe was holding the stick when he fell," Akono finished.

They had seen Pepe's facedown free fall through the long expanse of open air, before trees rising to lower heights obscured his landing. Pepe was younger, larger, and stronger than Jacky. It was natural for him to want to dominate and lead, but Jacky's gentleness and calm demeanor had won him the support of the whole group. Nama, Dorothy, and the juveniles had all established good relationships with Pepe, but they would never take his side against Jacky. Even Becky couldn't go against Jacky. Without support from the group, Pepe couldn't beat Jacky in a battle on the ground, but he couldn't quell his hunger for power. Finally, he had tried aerial combat.

The caregivers had been calling to Pepe since he fell. They had heard one bark of fear soon after he landed, and nothing since. Now all the other eleven chimpanzees in the group were inside the satellite cage, having responded to the caregivers' beckoning drumbeat, and I had to go into the enclosure to find Pepe.

The anger Pepe had exhibited toward me during my pregnancy had dissipated after Annarose was born. His displays against me had stopped, and he had begun sweetly soliciting my interaction again. Unfortunately, I had remained somewhat nervous about approaching him for months longer, and as a new mother with a whole project to run, I had been too occupied to have a lot of time to spend with him. He and I had begun reestablishing some of our old rapport only a couple of months earlier. In any case, not knowing how badly Pepe was hurt, it was scary to enter the forested enclosure, as we all knew he could be volatile and might react to us being in his territory. I loaded a dart of anesthesia in my dart gun to carry along just in case and asked the caregivers if one of them would come with me. When Akono volunteered to go, I was grateful for his bravery.

After only a few minutes of searching, we found Pepe lying facedown on the forest floor, in the spot where he had fallen. Kneeling beside him, I saw that he was breathing. When I touched him and called his name, he spoke a soft fear bark, telling me something was wrong. Something was terribly wrong. Pepe was conscious, but he couldn't move.

"Go get a stretcher and some help," I told Akono. "We need to carry him back to the cage." Waiting for Akono to return, I bent low to the ground in front of Pepe's face, and through tears, groomed his face and tried to assure him that I would help him.

As darkness approached, Akono returned with Assou, carrying a bamboo bench we would use as a stretcher. When we turned Pepe over, he cried out in pain, and again when we lifted him onto the stretcher. I knew that if he had fractured a bone in his neck or back we shouldn't be moving him like this, but I couldn't leave him on the forest floor with night approaching. There were too many dangers. Other large mammals couldn't get through the fence into the enclosure, but carnivorous ants and, to a lesser extent, snakes were big concerns. We transported Pepe to an empty chamber of the satellite cage that had been vacated by Dorothy and Nama and lifted him as gently as possible onto a bed of woodchips and leaves that Agnes had prepared on the floor.

Pepe couldn't sit up or even lift his arms or legs, but lying on his back, he could move his head from side to side. I placed a kerosene lamp beside the cage, and with a big syringe I squirted water into his mouth. After swallowing gratefully, he opened his mouth again and again for more. I peeled bananas and a papaya and cut them into small pieces. When he ate them willingly, I knew he had a will to survive this devastating injury. I would have stayed with him that night and many others to come, but baby Annarose needed me in camp. She was with Helene from the village all day, but I needed to be with her at night. I posted our night watchman in the cage with Pepe and asked the volunteers to check on him throughout the night.

During the next week, I consulted with two neurologists, and from Pepe's symptoms, we concluded that he injured his spinal cord in the lower part of his neck. We had no way to take an X-ray to determine if he had fractured a vertebra, but the neurologists thought he might possibly recover if given enough time. The caregivers, volunteers, and I hand-fed and cleaned him and turned him frequently to prevent pressure sores. I spent hours with him every day, comforting him, doing gentle physical therapy, reflecting on the time I had known him, and praying, in my way, that he would recover.

Very slowly, for five weeks after his terrible fall, Pepe incrementally regained some ability to move his arms and legs. Although he was frustrated, he ate well and enjoyed the grooming and attention I and the others gave him. During these weeks, he was never able to sit up, but he began to vocalize with enthusiasm to greet the chimpanzees from his group when they visited his cage from the forest. I knew that Pepe would never be the same, but I was hopeful that he might regain enough functional mobility to have a decent life.

Then one night, the guard who was assigned to stay inside Pepe's cage to assure his safety abandoned his post. The guard's most important job was to watch for protein-eating ants and to soak the ground with premixed kerosene and water to change their direction if they should approach the cage. The villagers knew how to effectively turn away ants in this manner long before I arrived in the Mbargue Forest. In fact, they had taught me how to do it. But on this awful night while Pepe was completely defenseless against them, millions of carnivorous ants attacked him. At six thirty in the morning, Agnes woke me up banging on my cabin door.

She spoke through the door. "Sheri, Pepe is covered in ants. It's horrible!!"

It was the worst thing I could have heard. I jumped from my bed, and in my pajamas with sixteen-month-old Annarose bouncing roughly on my hip, I raced down the trail to the enclosure where Pepe was lying on a bed of woodchips. As I approached the cage, but before I could really see Pepe, the horror of what was happening hit me like a punch in the gut. I could hear my heart pounding in my ears.

"Goddammit!" The guttural profanity sounded like it came from someone else, but I vaguely knew it was my own scream of anguish. People were just standing there! Pepe was covered with ants! I saw his face. He was biting his lip.

"Get the woodchips out." I don't know if I actually said it out loud this first time, or if it was just the thought forming. Then I yelled it, *"Get the woodchips out!"* It was the only way we could rescue Pepe. We had to remove the woodchips with the millions of ants crawling through and then pick the ants off him. *"Get the fucking woodchips out!"* My tone was murderous. Volunteer Jennifer Schneider pulled Annarose, who was crying, from my arms, and for once, Annarose didn't resist leaving me. Agnes, another volunteer named Phillip, and I used our hands, our feet, and the small brooms normally used for cage cleaning to sweep away the ant-infested woodchips from around and under Pepe, whose pain and anguish were only reflected in the movements of his mouth when he bit his lip or ground his teeth. We doused our legs and hands with kerosene to repel the ants and still frantically slapped at them on our bodies as we worked, wincing at their bites, fighting the urge to run to safety. We swept the woodchips onto scraps of plywood, our makeshift dustpans, in order to move them outside the cage to a growing pile, which we repeatedly doused with kerosene. When caregivers Assou and Akono showed up for their 7:00 A.M. shift, we were able to work faster.

When the woodchips were gone, we spent the next several hours painstakingly pulling thousands of the hideous flesh-eating ants from Pepe's nose, eyes, ears, and everywhere else on his body. They were even inside his anus, and I used my fingers to dig them out. I spoke to Pepe constantly as I worked to save him, but he didn't budge, didn't look at me, didn't respond in any way to my attempts to comfort him. He seemed to be somewhere else, keeping his eyes straight ahead toward the roof of his cage. I imagined him trying futilely to save himself with his limited arm and leg movements, until complete exhaustion caused his surrender.

The next day, we built a wooden bed off the ground and put the legs in buckets of kerosene so the ants couldn't climb up. I cursed myself over and over for not doing it sooner, but we hadn't wanted to risk causing damage by moving him. The day after the attack, Pepe started moving one of his arms again, but he was no longer moving his legs. Even before the attack, we hadn't been completely successful in preventing pressure sores. Now the ants had caused new wounds that were limiting the positions in which he could comfortably lie. Broad-spectrum antibiotics weren't controlling the infection, and analgesics weren't controlling the pain. Pepe stopped eating, and one morning he turned his head away again and again when I tried to give him water. When his eyes met mine, all I saw in them was suffering. I considered starting intravenous fluids, but I no longer had hope for Pepe recovering. He was giving up, and I needed to let him go. A few days after the ant attack, all I could do was help him die.

Once I decided to end Pepe's life, I felt an urgency to do it quickly, to not let him suffer another minute. Akono was with me at the cage, and I knew I needed to take the time to explain my decision to him and Assou, Pepe's main caregivers. They had seen much of suffering in the villages, but they had never seen or even considered euthanasia. The idea wasn't part of their culture, and I didn't know how they would react. I asked Akono to run and collect Assou and Agnes as quickly as he could, while I said my goodbye to Pepe. I caressed Pepe's hand and held it to my tear-streaked cheek. Even the day before, with my hands supporting his elbow, Pepe had used his fingers to groom my face. Now, through all his pain, he didn't try. "I'll make it end, my darling friend. It'll be over soon." After I explained to Agnes, who already knew I might make this final decision, she found the best words to explain to the caregivers. Akono cried, but neither he nor Assou tried to change my mind. I asked Agnes to explain to the other volunteers, too, so they wouldn't learn about it after Pepe was gone. Free to move forward, I ran to my little veterinary clinic, drew up the thick, pink euthanasia solution, and raced back to Pepe's side. I was no longer crying. I had to do this last thing for Pepe to the best of my ability without hurting him unnecessarily or upsetting him. I looked into his eyes one final time but fought my impulse to hug him, as it would have only been for me and might have caused him more pain. I asked Akono to groom Pepe's face and speak softly to him while I injected the euthanasia solution into his vein. By the time I looked from the injection site to Pepe's face, he was gone. My grief was suddenly overwhelming, but it was mixed with a sense of relief that this beautiful chimpanzee who had taught me so much wasn't suffering anymore.

All the chimpanzees who had lived with Pepe were still in their satellite cage when we wheeled his body past it that morning. They all came to the front of the cage, and I realized that somehow they knew immediately that he was dead. There were no loud barks of protest to us taking him out of the cage, as I might have expected. They met us mostly with silence, except for a single mournful cry from Becky. Her particular gut-wrenching vocalization, like none I had heard from any of them before, was an unmistakable expression of grief over the loss of her friend, her "brother." A gardener I had hired from the village was cutting grass with his machete a few yards from the cage. He hadn't been around the chimpanzees much, but hearing Becky's cry, he groaned reflexively and shook his head in sympathy.

In weeks to come, I blamed everyone for Pepe's suffering—the volunteers who knew the ants were bad that night, the irresponsible little shit of a caregiver who left his post, and most of all myself for not protecting Pepe. All my fury was channeled into killing ants—not in revenge exactly, but to protect the chimpanzees from further attacks. I shipped in several different ant poisons from the United States and spent weeks looking for ant nests. The ants moved in narrow columns during the day, and at night, they spread out hunting for animal protein. It was a bad year for ants, and I found these columns every day around the enclosures or on the road when I was driving. I coated the moving ants with poison of one kind or another, always hoping they would take it back to their nests, to their queen. Agnes told me, correctly I would later learn, that protein-eating ants were an important part of the ecosystem—controlling other insects by eating larvae, and controlling diseases by eating dead animals from the forest floor. If I were to succeed in killing them all, it would likely be an ecological disaster. I told her I would take my chances with whatever other insects or plagues descended upon us.

Certainly, logic was not my master during those sad weeks after Pepe's death. I didn't care about being responsible, and I didn't care what anyone thought. I'm pretty sure the villagers thought I had gone mad, and I probably had, temporarily. I was heartbroken and guilty and pissed off and scared the ants would get another chimpanzee. It was probably fortunate that my ant-killing rampage was indeed a fool's mission. The nomadic ants were endless.

Although I told people outside of Africa about Pepe's fall, and attributed his death to it, several years would pass before I could speak of the ant attack. It was too terrible to discuss; I was too ashamed that I let it happen. Even years later, when I could speak of it, casual questions about "the chimp who died by ants" cut me like a knife. How could people speak so lightly of the unspeakable thing that happened to Pepe? The suffering in his eyes, trusting me to give him relief in the final moments of his life, haunts me still.

sixteen

HEROES

By mid-2003, Jacky and Nama had been leading their small group of chimpanzees for almost three years. Each morning when the group left the satellite cage after breakfast, before disappearing inside the forest, they typically walked around the perimeter of their enclosure with Jacky in the lead, followed by Nama. On the opposite side of the fence, I loved following along with them.

Then I began to notice that Nama, not Jacky, was often taking the lead position. Jacky was following closely behind her, and if he found himself in front, he stopped and waited for Nama to go first. The caregivers began to laugh that Nama was wearing the pants in Jacky's family. I was confused about his odd behavior, until one day I figured out the reason for it. Like the other chimpanzees, Jacky loved his daily chewable vitamin, and he was always happy to pop it in his mouth after I handed it to him. When he accidentally dropped the dime-size, dinosaur-shaped orange tablet one bright July morning, it landed on the floor of his satellite cage. It was clearly visible to me, but I realized that Jacky couldn't see it. He eventually located it, but only after fifteen seconds of skimming around the ground with his hand. During an eye examination under anesthesia, I diagnosed cataracts, which were severely compromising Jacky's vision. Looking through my ophthalmoscope, I could barely see the retinas behind the cloudy lenses of his eyes. I believed he could still see light, shadows, and movement, but little detail.

Immediately, with Edmund's help back in the United States, I began searching for an ophthalmic surgeon to come to Sanaga-Yong Chimpanzee Rescue Center and remove the cataracts, but I knew the search would take time. For Jacky's safety I would need to keep him inside the cage until I could find help to restore his vision. The next morning when we let his group out, but closed the sliding door with him inside, he banged the door furiously for over an hour and then fell on his back on the floor in a screaming tantrum. That Jacky's capacity for leadership had survived decades of lonely isolation in a small cage was nothing short of miraculous. Given the opportunity, he had become a dignified alpha male—so different from the pitifully neurotic chimpanzee, the "mad chimpanzee," I had met at the Atlantic Beach Hotel. To see him writhing and screaming in anguish on the floor of his cage broke my heart and challenged my resolve to keep him inside. Up to this point my relationship with Jacky had been defined by respect. To a large extent since we brought him to the sanctuary, he had been able to make his own decisions about where to go and when. He had acquiesced freely to our routine of going out to the forest during the day and coming in at night. To force him now to stay in the cage when he objected to it so strongly and emotionally seemed terribly wrong, but I couldn't knowingly risk his life. In the following weeks I kept Jacky inside the cage, but I almost always kept Becky or Caroline, or occasionally Dorothy and Bouboule, inside with him so he wasn't alone. With no alpha male, Nama became the leader of the small group so I couldn't keep her inside with

Jacky. As the weeks stretched to months, Jacky went completely blind.

In the late fall of that year, my dear friend Susan Labhard, an advanced-practice nurse I met when she was a client at my veterinary clinic in Portland, Oregon, was aware of my effort to restore Jacky's eyesight when she traveled to San Diego for her Navy Reserve service. On my behalf, she knocked on the door of the navy's ophthalmic surgeon Dr. Jim Tidwell, and after a brief introduction to the work we were doing at Sanaga-Yong Chimpanzee Rescue Center, she asked him if he would be willing to travel there to perform surgery on a chimpanzee. Dr. Tidwell had performed many cataract surgeries in remote areas around the world as part of the navy's humanitarian medical service. He liked new challenges and new places. The idea of performing what was probably the first cataract surgery in a chimpanzee, and most certainly was the first in a chimpanzee in Africa, held some interest for him, quite apart from his desire to help a chimpanzee.

"It was something different," he told me later, as an explanation of his initial interest.

In January 2004, we paid Dr. Tidwell's airfare and he took two weeks of his vacation from the navy to travel with his expensive surgical microscope from San Diego all the way to the Mbargue Forest of Cameroon. His goal was to remove Jacky's lenses, rendered completely opaque by cataracts, and replace them with transparent lens implants made for a human. He had worked in all sorts of difficult circumstances and wasn't daunted in the slightest by our working or living conditions at Sanaga-Yong Center. I was prepared to coddle the California doctor in every way I could possibly manage within the confines of our rustic camp, but Jim was easy to accommodate. He fit right in with us from day one.

In case something was to go wrong that might be prevented another time, Jim opted to perform surgery on one eye at a time, with a week between the two procedures. Although we would soon install electricity in the veterinary clinic, we didn't have it then. To run Dr. Tidwell's microscope, we rigged up a solar panel and battery on the ground just outside the clinic. It worked perfectly. After the first surgery, mine was the first face Jacky saw with his new good eye. His gaze moved from my face to the forest outside his cage and back to my face again. His spark of recognition was unmistakable. Jacky could see again!

During the week between Jacky's first surgery and his second, Jim and I made a tour of the villages, performing eye exams. He had brought enough lens implants to perform surgeries on three people besides Jacky. We found seven elderly people with cataracts, so he had to determine which three people needed surgery most urgently. It was disturbing to realize that the people we didn't choose might never have the opportunity for surgery, but at least we could restore the sight of a few. We scheduled the surgeries of two women and a man, who were almost completely blind, in our small, one-room veterinary clinic.

Madame Jacqueline was the courageous first patient who allowed herself to be led to our clinic by her nephew Mvoku Samuel, who was working as a caregiver for a group of our baby chimpanzees. The day after the surgery, we went to the village of Bikol 1, where inside Madame Jacqueline's tiny house, leaving the door open for sunlight to enter, we took off her bandages.

When Jim asked if she could see, she joyfully exclaimed, "I see my pot!" She was referring to her cooking pot on the dirt floor of the room that had very little else in it. In this impoverished community where most people had no things, Madame Jacqueline's pot was something she treasured and almost surely hadn't been able to use in a long time. She would be cooking again soon.

Madame Awa from the village of Mbinang was the second person to have the surgery, and hers, too, was a success. Sadly, the very elderly Mr. Theodore from the village of Meyene didn't show up for his appointment. Like other people of his age in the village, he didn't read, and he spoke only Bamvéle, his tribal dialect. If he had ever heard a radio or seen a television, he could not have understood the spoken words. It wasn't so surprising that he hadn't been able to trust what we offered, but we were disappointed for him and sorry we didn't have time to schedule someone else who needed the surgery. For several years afterward when I saw Mr. Theodore being led around in the village, I wondered if we might have said or done something else to inspire his confidence.

Jim also brought some reading glasses to give out to people who had become far-sighted with age, and many people were delighted to receive them. One man who got a pair of Jim's glasses was Chief Bernard of the village of Meyene. After determining that the chief was having problems seeing objects that were close to him, Jim showed him a page in an open book and asked him if he could see it clearly, anticipating his negative reply. Afterward, Jim handed the chief a pair of glasses and told him to put them on.

"Can you read it now?" he asked the chief.

"No," the chief replied.

Jim chose a stronger pair. When the chief had them perched on his nose, peering through them at the page of print, Jim asked again, "Can you read it now?"

"No," the chief replied again.

Puzzled, Jim asked, "Do you see it better than you did before?"

"Yes," Chief Bernard replied. "I see it very well now, but I still can't read it."

Jim and I cracked up at the chief's deadpan joke, and through his own chuckling he said he would be happy to keep the second pair of glasses.

After Jacky's second surgery, his vision seemed perfect. When I placed his vitamin or peanuts on the floor in front of him, he could see them and pick them up instantly. The juvenile boys who had been sneaking some of Jacky's fruits and nuts for several months were quite surprised when his accurate fist and vocal rebuke delivered the somewhat painful and clear message that henceforth he would be keeping all his food. He went back into the forest, and life for him and his group returned to the way it had been before, with him leading.

By April 2004, we had completed a big, new enclosure complex where we would have a lot more space to integrate Jacky's group of older chimps with juveniles from our nursery. This new enclosure encompassed twenty acres of forest, and its satellite cage was four times the size of the first one we had built. One at a time, we moved the eleven chimpanzees in Jacky's group the three hundred yards from the old enclosure to the new. The four adults—Dorothy, Becky, Nama, Jacky, in that order—each voluntarily entered a transport cage when I asked and allowed themselves to be carried over to the new enclosure. The caregivers and I were able to carry over in our arms five of the maturing juveniles, who were by then between five and seven years old and hadn't been held by a human in over two years.

We had to give light doses of anesthesia to only Caroline and Mado. When I entered the cage to carry Caroline, now eight years old, she hugged me sweetly, but I suddenly knew she wouldn't be carried over easily. She would insist on going back to camp, messing around with the solar system, and there was no predicting what other mischief. I didn't have time for it. I injected her quickly by hand when she wasn't looking and then tried to fool her by looking around the air and acting fearful like a nasty bee was in our midst. I doubt if she believed me, but she turned her attention to her discomfort at the injection site. I sympathized and said I was sorry by grooming around it. Six-year-old Mado was more difficult to read. Her ordeal in a Cameroon biomedical laboratory before she came to us had destroyed her ability to trust humans. When I tried to carry her over, she panicked as I walked out the door of the cage and bit me several times during her forceful descent from my arms. She chose to return to the cage rather than flee into the forest. I was left with nasty bruises on my face and arm, but nothing more serious. In the end, I had to anesthetize this frightened little prepubescent girl with a blow dart in order to move her. So as not to traumatize her more than necessary, I crept to her cage in darkness, while she was sleeping, and blew a dart of anesthesia into the muscle of her thigh. Agnes Souchal worked with me late in the night to unite Mado with her group.

In the new enclosure we integrated other juveniles, eventually expanding the group to twenty-six under Jacky and Nama's watch. Conflicts among the maturing juveniles never escalated, and younger, smaller babies were never in danger with the two of them around. We introduced babies as young as two years to the group, and if ever a bigger juvenile or teenager played too roughly, Jacky and/or Nama could sort it out with little effort. From a seated position, Jacky with his hair on end could interject with a single bark, as if to say, "Don't make me get up!" That was usually all it took.

Older juveniles often referred to Jacky with frequent glances in his direction to make sure their behavior was acceptable. Less respectful conduct could elicit harsher discipline. When Jacky puffed up and stomped over to the transgressor, sometimes actually stomping *on* him, screams of fear were soon followed by apologies to Jacky. Even though Jacky could get tough, his gentleness was most striking. When Mado gave birth to Njabeya in 2007 (my tardiness in performing a vasectomy on Moabi resulted in the birth of this baby girl whose name means "gift" or "treasure" in one of Cameroon's languages), she trusted Jacky to touch her precious newborn. To him, Mado bestowed the privilege of holding Njabeya's foot. Tenderly cradling the tiny, soft appendage in one of his huge hands, Jacky fingered the minuscule toes with his other, an expression of awe on his face.

Most of Jacky's acknowledgments of me were subtle. When I approached his big satellite cage in the early mornings after breakfast, whether he was lounging in a grass nest on the floor of the cage or dangling his legs from a platform above my head, he acknowledged me by extending his foot in my direction, often without bothering to actually look at me. When I extended my hand to grasp his foot, always grateful for his acknowledgment, he invariably squeezed my hand affectionately, making good use of his opposable big toe (a useful anatomic feature, our loss of which is lamentable). This "handshake" was our hello. I felt that Jacky delivered as much warmth with the minimal gesture as another chimpanzee might bring with a big hug.

One evening just after my return from a two-month visit to the United States, I crouched near Jacky's satellite cage as dusk was approaching.

"Hey, Jack," I called. "It's me." He was about to ascend to a high platform carrying an armload of grass to make a nest for the night. Instead, he walked over to my side of the cage and extended his hand out through the bars toward me. I closed the space between us, somewhat tentatively since I had been gone awhile. While Jacky was kind and gentle with chimpanzees and he and I had a positive history, I couldn't forget that he could be violent with humans. When his eyes moved to my hairline and his mouth tightened in concentration, I knew he wanted to groom me. He came in peace, the only way Jacky had ever come to me. I scooted close to him and bowed my head so his fingers would have easy access to my hairline. For several minutes, I enjoyed one of the few times Jacky has ever groomed me. When I raised my head to look at his face, I found his eyes seeking mine. Now when he extended both his hands outside the cage toward me, I placed my hands in his, and we both squeezed. In the moments that followed, looking deeply into Jacky's eyes, holding his hands, I didn't speak, but the song in my head was my spontaneous and heartfelt message of respect to him. *I know who you are. I know who you are. I know the strong character that is yours.* Jacky understood my message, and he was telling me the same thing. We were two people communicating with body language we both understood.

The perimeter of the new enclosure we built for Jacky and Nama's group was about three-quarters of a mile. I still enjoyed accompanying them on their morning patrols, and though I stayed on my side of the fence, Jacky seemed to consider me a member of his group during our walks. If I fell behind, he stopped the group to wait for me. One morning, as I paused at a corner of the fence line to film the group walking ahead of me, I captured Jacky turning to use a beckoning hand gesture to request I hurry along.

Although I have been pleased and honored by Jacky's attention when he has chosen to give it to me, my deep admiration and adoration of him stem from my observances of him with the other chimpanzees. He earned my deepest respect for the venerable chimpanzee he was with them. But Jacky wouldn't have reached his potential as a gentle leader without the unwavering support and guidance of wise and brave Nama. Her loyalty, sense of justice, and courage were unsurpassed. Whether in collaboration with Jacky or acting alone, she always made admirable choices.

One of our former volunteers is probably alive today only because Nama put herself on the line to protect her. When I was in Yaoundé, we had a terrible accident at Sanaga-Yong Chimpanzee Rescue Center. One of our caregivers mistakenly let some of the chimpanzees into the human corridor that runs down the center of the large satellite cage between the two rows of chimpanzee chambers. From this protected corridor, caregivers and volunteers could give food, and medicine when necessary, to all the chimpanzees in Jacky's group. A volunteer was in the hallway alone when the mistake occurred.

On this horrible day that could have been her last, Kim (I'm not using her last name out of respect for her privacy) looked up to see huge teenager Bouboule, adopted son of Dorothy, charging down the hallway toward her. Finding himself in the corridor, somewhere he had never been, with a human he barely knew, Bouboule became extremely agitated and dangerous. If he had wanted to kill Kim, he could have done so within a few seconds. Fortunately, killing her wasn't his intention, but it could have been the tragic result he didn't intend. Pumped up by testosterone and trying to claim this new territory that had been off-limits, he used Kim as a display object as he raced back and forth down the fifteen-yard corridor. Over and over, with bone-breaking force, he stomped on the body of this petite young woman. If Kim had panicked, she probably would have died that day, but to her tremendous credit she went into an almost calm survival mode.

Nama was among the several other chimpanzees in the corridor with Kim and Bouboule. Curled in a fetal position, rendered barely conscious by Bouboule's latest pummeling and trying to protect her bruised and bleeding head between her arms, Kim noticed that Nama, sitting a few yards away, had puffed up uncomfortably. Kim didn't know Nama well, but when she saw Bouboule coming toward her again, she called out to her for help, knowing it might be her last few seconds of consciousness. "Whoo, whoo, whoo," Kim whimpered to Nama like a chimpanzee crying, and Nama responded to her plea. Not unlike how she had protected Bouboule's mother, Dorothy, from Pepe years earlier, she now intercepted Bouboule, who was much larger than she was, but ran from her nonetheless. Nama chased Bouboule away from Kim three times before the caregivers were able to dart him with anesthesia and get Kim out of the hall. Kim's whole body and face were left with deep bruises, her eyes were swollen shut, and doctors thought she had a concussion, but thanks to Nama's bravery and kindness, she survived with no physical wounds from which she didn't heal.

From the hospital, needing to recount what had happened as a form of healthy catharsis, Kim described to me the details of that awful fifteen minutes and what Nama had done to save her. In the early years when I spoke of Nama to the outside world, I often described her as lionhearted, but no lion's heart could be as courageous or kind as that of this alpha female chimpanzee. That chimpanzees can be horrifically violent is well documented. Popular media has publicized tragic cases of human victimization. That the chimpanzee species also includes heroes like Nama is less well known.

Until 2007, Jacky and Nama's group of chimpanzees still included Becky and Dorothy, the other original adults whose sad plights had inspired me to build Sanaga-Yong Chimpanzee Rescue Center. Bouboule's adolescent propensity for terrorism notwithstanding, it was an amazing group of chimpanzees, the leaders of which were living testament to the buoyancy of the chimpanzee spirit and their phenomenal capacity to recover. I came to think of it in grandiose terms as the *Jacky and Nama era,* because I so admired them as leaders and so enjoyed observing the peaceful society they created and maintained. When I was on-site at Sanaga-Yong Center, I was absorbed in it. Because our chimpanzee population had increased rapidly during the first few years, we had established three other social groups in other enclosures, which I also enjoyed, but observing and interacting with Jacky and Nama's group was special for me. I knew that I was living through a historic time that would hold significance in my life like none other. If only Pepe hadn't left us, it would have been perfect. I was immensely grateful and felt privileged to be a witness to the astonishing transformations of these extraordinary individuals. But my happiness stemmed from more than that. In order to bring these chimpanzees to this place and time, I had needed to find the best in myself. My satisfaction in it was complete.

seventeen

NECESSARY TRADE-OFFS

We advertised for volunteers on our website and on a couple of other sites that listed job opportunities for people interested in working with primates, and while I was at Sanaga-Yong Chimpanzee Rescue Center, applications often stacked up in my e-mail box. During my regular short trips to Yaoundé to visit George and take care of business that required an Internet connection, I conducted telephone interviews with volunteer candidates who seemed to meet our qualifications and made decisions about who we would accept. The volunteers were required to work six or seven days per week (depending on how many volunteers we had) preparing fruits and vegetables for the four chimpanzee meals per day, preparing and delivering milk formula for the babies three times per day and at midnight, taking babies into the forest when we were short of staff (especially on Saturdays), doing the marketing in Bélabo for the whole camp, and making sure the staff signed out all the equipment and supplies they needed each day. They were usually up by 6:00 A.M. and had little downtime, so I required people with stamina. I looked for people who could speak French at least on a conversational level, and I preferred people who had some experience living in Africa and/or working with chimpanzees or other primates. In reality, I rarely found anyone who met all the qualifications. Eventually, long-term American volunteer Karen Bachelder, who traveled to Cameroon and worked on-site at Sanaga-Yong Center many times, very capably took over the interview process, formalized it with strict reference requirements, and made recommendations to me about whom to choose. Before that I often made predictions about who would be able to adapt to our environment and serve the project well based on my gut reaction to them in the short telephone interview, as much as anything else. I got it right sometimes and wrong sometimes. I was always delighted when a good volunteer wanted to come back.

After volunteering for six months in 2002, Agnes Souchal came back to Sanaga-Yong Chimpanzee Rescue Center in 2004 with the intention of volunteering for another eight months. While she was on-site, she put the interests of the chimpanzees above everything else, certainly above concerns for her own comfort, or lack thereof, which was always my first measure of a good volunteer. A sure way for a volunteer to get on the wrong side of me was to complain about the facilities, which—even without running water or electricity—were luxurious compared to the tent where I started and contrasted with the surrounding villages, where people raised families in crowded huts on dirt floors. I appreciated that Agnes had no irrational fears of insects, spiders, rats, or even snakes. Her ability to adapt to the African forest probably surpassed even mine. To top off her good points, her subtle sense of humor made me laugh.

This second time Agnes volunteered, I worked more closely with her on chimpanzee care and staff management than I had the first time she volunteered. We had our cultural differences, but I thought she was nicely diplomatic for a French person, and she thought I was refreshingly frank for an American. She was restrained in her interactions with the employees, and as a result, she wasn't extremely popular with them the way some other more gregarious volunteers had been. But she tracked their individual problems, and when the two of us were alone together she advocated quietly for them as no one else had. Emmanuel needed help sending the orphans of his deceased sister to school. Albertine's daughter wasn't getting enough protein. Surely we should help them. I sometimes teased her for being too French in her notions of labor management, but I too understood that we were usually the last resort for our employees and the village people. If we didn't help them, no one would. We found common ground on staff-related issues, and we were in sync when it came to chimpanzee care. At the end of her second volunteer stint, when she had spent a total of fourteen months working seven days a week at the center, I offered Agnes the position of general manager, and she accepted it. A petite thirty-one-year-old Frenchwoman who had lived all her life in Paris, Agnes wasn't an obvious choice to become manager of a chimpanzee sanctuary in rural Africa, but I certainly wasn't an obvious founder of one either. Gradually, we became irrevocably bonded by our love for the chimpanzees in our care, and our working relationship evolved into a deep friendship based on the concerns that consumed us both.

In that same year of 2004, George Muna introduced me to Raymond Tchimisso, who was interested in collaborating on a sensitization campaign in Cameroon's West Province. During his travels around the town of Bankim as the personnel manager for a Chinese road-building company, Raymond had been touched by the suffering of strictly confined, captive chimpanzees. He had an idea about how to stop the killing and orphaning of chimpanzees in the Bankim area, where respect for the traditional chief was very strong.

Raymond's English was no better than my French, so our discussions about personal motivations were limited, but he made me understand that his first cousin was the sultan of Bankim, the traditional chief of a large territory, and that the relationship could be of use in a sensitization campaign to help chimpanzees. By coincidence, George was an old friend of the sultan, and as such he too had some friendly influence with him. Through a series of meetings between Raymond, George, and me, and others between Raymond, George, and the sultan, we planned and implemented a community meeting in Bankim.

Whereas the national laws against killing and capturing chimpanzees weren't adequately enforced, we hoped that an edict against it from the sultan would hold sway in his territory. To announce the momentous meeting, the sultan beat the drum literally and figuratively. The beating of an actual drum throughout the night in his palace summoned a team of messengers, which the sultan then launched on a walk through surrounding neighborhoods shouting the announcement of a meeting at his palace. "Elephant has sent me to call you to his palace to receive important information!" This was the cry of the messengers, and people understood that the meeting was compulsory. For reasons that were never completely clear to me, it was not considered respectful enough to refer to the sultan as "the sultan," so instead he was called Elephant, which was meant to symbolize his power and strength. At the same time the foot messengers embarked, Elephant sent motorcycle taxis to the more distant neighborhoods and villages to notify all the chiefs under his authority about the meeting, so that they in turn could inform the people in their smaller territories. Because I was occupied at Sanaga-Yong Center, I sent my short-term (as it turned out) administrative assistant Jojo with Raymond to the meeting, and they both reported on it to me later.

During the meeting, attended by hundreds of people, Raymond spoke about the problems chimpanzees were facing, and about the laws designed to protect them. "Our people of the Tikar tribe have been good warriors and good hunters. Until recently, we didn't have a government outside of our tribe, but now there is a national government, which has its laws. We are here at this meeting because Elephant has called us. If we had failed to respond to the summons, his majesty could have fined us. The national government's law says we can't kill chimpanzees, and if we do, they can fine us, or even send us to prison, so we must respect the law." Raymond tried to explain it all in terms his people could understand well.

After he finished speaking, the sultan himself rose to say that the killing and capturing of chimpanzees would now be prohibited in his territory. Afterward, the meeting attendees were treated to a feast of food and drink, funded by IDA-Africa, which was essential for inspiring goodwill in the community. There were a thousand other communities that would continue hunting chimpanzees, but we made some headway in this one with a single meeting. Maybe some chimpanzee lives were spared as a result.

Soon after the meeting, Raymond Tchimisso accepted the job of personnel manager and community liaison for Sanaga-Yong Chimpanzee Rescue Center. He brought his personnel experience, cultural wisdom, and diplomatic skills to our operation, while Agnes adeptly managed general operations and provided loving and astutely perceptive care to the chimpanzees. She didn't have a medical background, but her keen interest assured she was a quick learner for all things medical. Soon she was able to provide basic medical care for the humans and chimpanzees, and her need to consult with me about the details of every case decreased as she gained experience. Raymond's arrival to support Agnes and her continually increasing skill afforded me more freedom to be away from the site.

This freedom to come and go had become important to me as Annarose had reached preschool age, when I began to realize that my little daughter needed a different social and linguistic environment than I could provide her in the Mbargue Forest. Our cook, Cathy, had been bringing her daughter Ilsa, the baby I had "delivered" in the back of our pickup truck, to play with Annarose once or twice a week. This was the only regular contact she was having with another child. Like most village children, until she would start school, Ilsa would learn to speak only Bamvéle. Annarose was picking up this local language both from her new nanny, Veronique, and from Ilsa. While our employees were delighted that Annarose was greeting them in their dialect, I was quite sure that this language spoken by only a handful of African villagers wouldn't serve her well in the world. I asked Veronique to speak French instead of Bamvéle to her, but it was hard to enforce and impossible for Veronique to understand the importance of it. Annarose's speaking of Bamvéle was a source of pride for Veronique. The only English my daughter was hearing was from me, and I was busy all day almost every day. My time with her was limited to lunch breaks and a couple hours each night. In addition to my relatively minor concerns about the language, I was worried that the isolation wasn't healthy for Annarose. Although she seemed like an extremely outgoing, self-confident three-and-a-half-year-old, perfect in every way in her mother's eyes, I thought she needed exposure to other children and more academic stimulation than I could give her. In 2005, I enrolled Annarose in preschool in the city of Yaoundé. With both Agnes and Raymond managing at the center, I could spend much of my time in the city with her. I missed the chimpanzees terribly and didn't care for Yaoundé, but I was compelled to be with my child and provide the best life possible for her. The chimpanzees would always require my dedication and support, but at this time Annarose needed my consistent presence more than they did. Barring any emergencies that took me there more often, I was spending an average of one week per month at the center. When I was there, Annarose stayed with her dad in the city.

George and Annarose had a sweet relationship. His hands-on participation in parenting went against the cultural norms for men of Cameroon. Although we had been together only sporadically, he had changed her diapers and fed her from the beginning and was proud of his involvement in her care. When Annarose was about a year old, I heard George speaking to his brothers who had experienced fatherhood in a more conventional way. "Men in Cameroon just don't know what they're missing. Taking care of Annarose, changing her diapers, with her little eyes looking into mine with all that trust, it's like nothing else." I sometimes thought he was more like a grandfather than a father, perhaps enjoying with Annarose the pleasures of fatherhood that he regretted missing in his children who were babies when he was much younger.

George and I never had a wedding ceremony, but with young Annarose, we shared an apartment in Yaoundé for five years. He introduced me as his wife, and I referred to him as my husband, but with much of his work in Cameroon's Southwest Province and much of mine in the Mbargue Forest, it seemed that one of us was always leaving. Although I longed for my daughter when I was in the forest, I was never really lonely there. I cherished my friendships with the chimpanzees (and increasingly with Agnes), and after days filled with physically and mentally demanding medical care or construction work, I usually fell into bed completely exhausted by 9:00 P.M. In contrast, in the Yaoundé apartment alone at night after Annarose was asleep, I often felt lonely. My days in the city were occupied with providing logistical support for the center, often gathering and sending supplies, or with attending various meetings in the ministry or with volunteer arrangements or with fund-raising efforts. I also had people to organize—my assistant, along with a housekeeper and eventually a driver paid by George. But my nights in Yaoundé were often long and empty. Nonetheless, I have some lovely memories of the times when George was there with me, particularly of his special brand of chivalry. He must have met a big challenge in wielding it for me—given the independent and headstrong kind of woman I was—but he managed it sometimes.

Early one evening, we were walking down a dimly lit road in Yaoundé when three men appeared suddenly from an adjoining alley a few yards ahead, walking at a fast clip toward us. With no time to think, George's instinctive reaction was to step in front of me with his arms extended and shout *"Hoa!"* Fortunately, the guys veered away from us and disappeared on the other side of the road. If it had come to fighting them off, I most likely could have held my own as well as George and would have certainly fought beside him, instead of hid behind him, but his brave gut impulse to protect me, potentially at his own expense, impressed me. Looking back, it still does.

One Sunday afternoon—I remember it was Sunday because the pharmacies were closed—George suffered alongside me while my fever was rising during a terrible case of falciparum malaria. Because my brain was responding to the infection and telling my body that its temperature should be much higher than it was, I was miserably cold and racked by uncontrollable shaking, which was exacerbating the intense pain in my joints and head. I was desperate to get warm and stop shaking. George turned off the fan that normally made the hot climate bearable, and, after wrapping me in all the blankets he could find in the house, sat beside me on the bed to hold me while I shook. At one point the housekeeper entered the sweltering room and cracked the balcony door (out of concern for George, I'm sure). The slight movement of warm air coming through the narrow opening at the door felt to me like an arctic blizzard. Through chattering teeth, incapable at that moment of considering anyone else, I moaned for her to shut the door, and she did. Finally, when my fever was high enough—103.8 degrees as it turned out—that I stopped shaking, I was aware of George sitting beside me on the bed, his clothes soaked completely through with sweat. He had suffered through a long suffocating sauna on my behalf, and I knew he hadn't left my side once to seek relief. I touched my fingers to his wet shirt.

"I'm sorry," I told him.

"You held yourself well," he complimented me as he held a cup of water to my mouth. I hope George remembers moments of kindness from me, too. I'm quite sure he remembers times I wasn't kind.

Despite the love and tender moments between George and me, domestic harmony was elusive for us. Although he helped with my work for the chimpanzees in many ways—serving on the board of IDA-Africa in Cameroon, providing contacts and trusted advice, lending some of his employees when I needed extra help with a project, contributing financially—he sometimes complained bitterly that I put my work before everyone and everything except Annarose. When he was angry, he accused me of putting it even before her. He couldn't understand that my devotion to the chimpanzees stemmed not only from my love for them, but also from my sense of responsibility to them, which really did come before everything except my responsibility for Annarose. Their survival depended on me. George's didn't. My husband certainly didn't appreciate my lack of skill in "running the house" in Yaoundé. I had a housekeeper to help, but I wasn't interested in coordinating the dinner menu a week in advance, as he thought I should. He thought it was my role as the woman to do this kind of planning for the household, and I shunned this imposed role outright, out of principle if nothing else.

In hindsight, it probably wouldn't have cost me much effort to "run the house" a bit more to his liking. George and I loved each other, but neither of us was willing to compromise enough to smooth out the stark cultural differences that were magnified by intimacy and close proximity. Eventually, we determined our frequent arguing hurt Annarose more than our living apart would do. But our mutual love for her, and our respect for each other, inspired us to remain close friends after we stopped sharing a household.

Back in the United States, Edmund was still the development director and "liaison officer" for IDA-Africa. He was the one I called to organize shipments of equipment or supplies that I couldn't get in Cameroon and to do Internet research on all kinds of topics, since my connection in Yaoundé was slow and unreliable. He worked with volunteers to organize fund-raising events in Seattle, Portland, and other cities, and I attended and spoke at them when I traveled to the United States with Annarose for about two months every summer. Edmund was a vital part of IDA-Africa for a decade. During this time, our deep and enduring friendship grew, and he became an important "family" member to Annarose. The two of them spent many hours together every summer, and they grew to love each other very much. Through it all, Edmund dated various women who usually didn't understand or appreciate his friendship with me, but eventually he found happiness in his marriage to Cindy Scheel, who was his perfect match and who became a close friend to me and Annarose, too.

eighteen

FAREWELL TO OUR SASSY GIRL

During Becky's first few years at Sanaga-Yong Chimpanzee Rescue Center, I joked that she, unlike other chimpanzees, was not unpredictable. On the contrary, her penchant for troublemaking was very reliable. On many occasions Becky delighted in making mischief that upset her caregivers and me.

One afternoon, she brought in a long stick from the forest and poked it between the bars of the metal mesh ceiling of her cage to tear a gaping hole in the zinc roof above it—the very roof that protected her and the others from rain while they slept. "She clearly didn't think this through," I joked with the caregivers. The more vigorously the caregivers and I scolded, the more enthusiastically she stabbed at the roof, pausing only for a moment to stick her tongue out at me. It was good fun for her. Only after we walked away and ignored her did the activity lose its appeal.

Another time, one of her caregivers made the terrible mistake of leaving a machete just outside her cage chamber. It was a temptation that few chimpanzees could be expected to resist, least of all Becky. Fortunately, she was in a cage chamber alone when she committed the heist, so the danger was less immediate than it would have been if she had been with other chimpanzees. The more nervous we became about the potential harm she might cause herself and the more we pleaded for a trade, the more gleefully she swung the machete around. She held it by its handle as she had seen the caregivers do, never touching herself with the blade. When she finally got bored with it, she showed some interest in negotiating a trade, but her price was high. Not for bananas or papaya or peanuts—foods that were frequently part of her diet—but only for a cup of yogurt, she finally handed the machete to me sweetly.

Over time, Becky's relations with both chimpanzees and humans became more gentle and accommodating. A few years after her arrival at the center, Becky actually assisted the caregivers by performing a helpful cleaning chore and was encouraged by our praise to continue. The caregivers routinely used small straw brooms to brush the spiderwebs and dust from the mesh of the cage walls, but they neglected to climb up to sweep the cage roof or the sections of the sixteen-foot mesh walls that were high above their heads. One day when caregiver Emmanuel accidentally let Becky get hold of a broom, she used it to diligently sweep the spiderwebs from those high parts of the cage that the caregivers couldn't reach. Emmanuel found me in camp and asked me to accompany him to the cage to see what Becky was doing. I found her knitting her brow in concentration, performing a useful task and seeming happy to please us all. I told the caregivers to let Becky have a broom whenever she wanted one. However, quite understandably from my point of view, she lost interest in the tedious task fairly quickly and didn't choose to do it many times.

Like Dorothy, Becky was an adoptive mother. After Pepe died, it was Becky alone who carried Gabby—the youngest and smallest in the first group of juveniles we integrated. She allowed him to sleep with her on her tire-bed, which we moved with her to the satellite cage of the new enclosure, and she even shared her food with him. As Gabby grew older and as we integrated younger chimpanzees into the group over several years, Becky nurtured other babies, including Luke, Lucy, Future, and Emma. It was moving to see Becky content in these tender matriarchal roles.

Perhaps the most striking change in Becky's behavior came in her relationship with Dorothy. One morning in 2006, I returned to Sanaga-Yong Center after being in Yaoundé for three weeks and eagerly joined the group for their morning three-quarter-mile stroll around their enclosure perimeter. By this time in her life, Dorothy was rarely going on patrols anymore, but when I got back to the cage that morning, caregiver Assou told me that Dorothy had followed us—going along, he was sure, because I was there—bringing up the rear. I hadn't even noticed her. I rushed to make the tour again, this time trying to catch up with Dorothy wherever she was along the fence line. When I found her, about one-third of a mile from where she had started, she wasn't alone as I had feared. Becky had stayed behind with her and now rested her hand supportively on Dorothy's shoulder. I squatted beside them, separated by a few feet and an electric fence line, to see if Dorothy was okay. She was covered with sweat and breathing heavily. I sat with them about ten minutes until Dorothy seemed rested, then stood and took a few steps.

"Come on, Dorothy. You can make it, old girl," I said, trying to persuade her. When Dorothy rose to walk, Becky hobbled along beside her, using only one hand on the ground and the other around Dorothy's upper arm, pulling her along. Becky and I let Dorothy set the pace, and the three of us made several long rest stops along the fence perimeter before we finally got back to the cage. It was the last time I saw Dorothy walk around the enclosure.

In the mornings and evenings, Dorothy and Becky were frequently together. In the heat of the day, Becky usually disappeared in the forest with the rest of the group while Dorothy sat in the shade at the edge, but whenever the group was all together, near or inside the satellite cage, any stranger watching would know that the two female chimpanzees were friends. What a change a few years had made!

In early 2007, I was on a visit to the United States while my friend Dr. Kerri Jackson, a talented veterinarian whom I trusted completely, was on-site at Sanaga-Yong Chimpanzee Rescue Center. Agnes called to tell me that Becky was very sick—pale, and showing signs of extreme pain in her abdomen—and that Kerri thought she needed to perform surgery immediately to find and hopefully treat the problem. From the symptoms they described, I agreed. During an abdominal exploratory, Kerri found the same kind of adhesions I had found in Becky's belly six years earlier. This time, Becky was bleeding profusely from somewhere, and it was difficult to discover the source through the extensive adhesions that glued all the contents of her abdomen together. Racing against time, trying to find the source of the bleeding before it was too late, Kerri painstakingly dissected through adhesions to finally discover that Becky was bleeding from a ruptured uterus that contained a five- or six-month-old fetus. Becky, not yet thirty years old, died during the surgery.

With that terrible undiagnosed pathology in her abdomen, Becky probably wasn't destined to live to an old age, but the pregnancy hastened her death. I was devastated, and I felt guilty. To prevent pregnancies in our chimpanzee residents, I had performed vasectomies on our adult males, using a technique I had learned from a human surgery book. I had been performing them on younger males as they approached puberty, but I obviously had waited too late for one of them. I was sure it was Moabi, who I had intended to vasectomize before I left for the United States but had run out of time. In the days following Becky's death, Agnes spent many hours collecting urine to run pregnancy tests on all our teenage (eight- to thirteen-year-olds) and adult female chimpanzees. This is how we discovered that nine-year-old Mado was also pregnant, and it was the time we started all the other pubescent and older females on birth control pills. Mado would give birth to the only chimpanzee who, as of this writing, has ever been born at Sanaga-Yong Center.

On the sad day that Becky died, I sat useless and grieving in the United States. I could do no more than write a tribute to her life, which we shared on our website with our supporters:

> When we met in 1997, your childhood and your adolescence already had been stolen from you. But your penetrating brown eyes stared at me from behind the bars of your tiny cell, and I saw that you were surviving and curious and hopeful in spite of all you had lost. Confined and bored out of your mind for so many years, somehow you were still vibrant. Becky, your eyes grabbed my heart, and they never let go . . . I knew that every single day counted for you, and that your fortitude deserved mine . . . dearest Becky—tough chick, sweet lady, flirt, goofy face-maker, lover of your comfortable "nest" made of an old tire where you slept for seven and a half years, overeater of bananas, surreptitious plotter, cleaner of cobwebs, "sister" of Pepe and Jacky, mourner of Pepe, best friend of Dorothy, frustrater of pubescent boys, adoptive mother of Gabby, matriarch and protector of little Luke, Lucy, Future, Emma, and others— during your years with us at Sanaga-Yong Center I know you had a rich life that you cherished. And oh how we cherished you!

Agnes, Raymond, and the caregivers buried Becky beside the enclosure where she had lived with Jacky, Nama, Dorothy, and more than twenty other chimpanzees. From the enclosure, Becky's family and friends watched them bury her body, and then all but Dorothy wandered away into the forest. For the rest of that sad day, and significant parts of many days to follow, Dorothy chose to sit directly across from the gravesite of her close friend Becky.

nineteen

DOROTHY'S LEGACY

For several years after she had adopted Bouboule, Dorothy joined the rest of the chimpanzees for the morning patrol around the fence line, and when they were in the satellite cage or at the edge of the forest, she enjoyed grooming and supportive relations with everyone. For a couple of years during the period when Bouboule refused to leave her side, she even went into the forest with the other chimpanzees. She couldn't, or wouldn't, climb trees, and I imagined her sitting calmly on the cool forest floor while Bouboule played around her or above her with the other juveniles. I hoped they dropped some of the delicious yellow fruits from the umbrella trees for her to enjoy. Of course, I didn't know what happened in the forest; I only knew what I saw outside of it.

As Bouboule became an adolescent and Dorothy grew older, she accompanied the group on their patrols less often. She usually rested in the shade at the edge of the forest or sometimes she found a reason for lingering just outside the cage. The cage was locked for cleaning until around eleven A.M., but sometimes Dorothy spotted leftover fruit on the floor that she could reach from outside. One morning, she spotted a half-eaten avocado, her favorite food, lying where it would have been within her reach had the holes in the mesh been large enough to accommodate her forearm. She pointed at it and grunted, looking me in the eye, asking me plainly to fetch it for her. I was in the hallway of the cage without a key to the chamber where the fruit lay. I pulled on the lock of the cage and showed her my empty hands, so she would understand that I couldn't enter the cage either. Grunting to express her mild annoyance at my uselessness, she turned away from me to solve her own problem. At a bush near the edge of the forest, she took her time to test several branches before breaking one off. Back at the cage, she used the freshly broken end of the quarter-inch-diameter stick as a spear to pass through a hole and stab the avocado. It took a few minutes of effort because the avocado kept falling off the spear, but Dorothy was able to use her cleverly chosen tool to bring the avocado within reach of her thin fingers. Grunting happily, she ate the green flesh of the fruit, then plopped the seed into her mouth to suck until it stained her tongue bright orange.

Dorothy was adept at communicating what she wanted people to do for her, including when she wanted someone to disappear. With a disgusted grunt and a back flip of her hand, she told people she didn't like to go away. I always considered it to be Dorothy's version of the middle finger gesture. Had she ever applied it to me, it would have hurt my feelings terribly.

Although she became more and more physically frail, Dorothy was powerful within her social group. She continued to defend Bouboule until he was a big teenager, larger than she was—until long after the only reason he needed defending was the mischief he made. One morning I was squatted beside the enclosure about a hundred yards from the satellite cage "chatting" with some of the chimps, when a loud conflict suddenly erupted just a few feet from me. I stood to see what was happening, but it was hard to take it all in. At least fifteen screaming chimpanzees were involved, and I soon realized that Dorothy was right in the center of the fracas, but it was incomprehensible that so many in the group would be angry with her. A second or two later, I realized that Bouboule, then the same size as Dorothy and much more muscular, was quite absurdly hiding and screaming behind her. She was barking and throwing punches in his defense. It ended quickly because no one wanted to go through Dorothy to get to Bouboule, although in all honesty he probably deserved whatever the group wanted to dish out to him.

As much as I loved Dorothy, I didn't like her son much at that stage of his life. I had known him since Estelle and I had rescued him as a two-year-old from a village where he had been tied up in ankle-deep mud beside a pig, malnourished and covered with sarcoptic mange—arguably the most itchy skin disease known to man or chimp. I had bottle-fed him and cared for him when he was ill. I had watched him mature as Dorothy's love had seemed to transform him from an insecure, lonely baby to a happy, self-confident chimp child, full of laughter. Unfortunately, as he approached puberty, his mean streak surfaced. Almost all adolescent male chimps can be unpredictable and dangerous. They wield phenomenal physical strength for their size and are usually eager to prove themselves, gain power, and climb the social ladder. Bullying is a common chimpanzee tactic for trying to demonstrate dominance, but Bouboule had a particular penchant for it that our other male chimpanzees his age did not.

Dorothy used her back fist punch to defend Bouboule many times, but she occasionally used it to discipline him as well. One hot afternoon when she was inside enjoying the cool of the shaded cage, I sat just outside the cage enjoying her company. She was grooming my face when Bouboule, who I believed was jealous, ran in from the forest to surprise me with a painful punch to my nose. He had backed away from the cage wall and was coming back for a second go at me—although this time it was merely intimidation because he could see I was cupping my aching nose well out of his reach—when Dorothy's back fist, delivered with perfect timing, landed squarely on his chest. Accompanied by a harsh vocal scolding from Dorothy and from me, for whatever my input was worth, it sent Bouboule running back to the forest.

Another incident about the same time, during the last month of her nine-month pregnancy, involved Mado. Mado had been one of those first six juveniles introduced to Dorothy and the other adults in January 2002. Now Mado was heavy and easily tired, and by her own choice, she spent more time reclining in the satellite cage and less in the forest. One afternoon, she was resting on the cool concrete when Bouboule began menacing her hatefully—running past her over and over, slapping her hard on the back each time. All the other chimps were deep in the forest, but from her shady spot at the edge of it, Dorothy heard Mado screaming in distress. She rushed, as much as Dorothy was able to rush, to Mado's rescue, firing off a barrage of barking as she entered the cage, quickly followed by a back fist punch at Bouboule that didn't land well. That the punch failed to land didn't matter. Dorothy's scolding was enough. As Bouboule stomped out of the cage, Dorothy accepted Mado's grateful embrace with open arms, patting the back of the much younger chimp soothingly as she held her. Watching from the corridor of the cage, it felt like my love for Dorothy was too much for my heart to hold. Somehow it was painful, and I cried.

In the late winter of 2008, we had an outbreak of a terrible respiratory infection that spread like wildfire through all our chimpanzee groups. Dorothy, like many of the others, became very ill. Unlike the others, she didn't bounce back within a few days. Although she began to enjoy her food again, she was losing weight in spite of it, and her respiratory rate remained rapid. While I was in Yaoundé, Agnes entered her cage to collect blood for some tests I requested, and Dorothy, who was happy for the close contact with Agnes, cooperated fully for the procedure. But the blood work was unremarkable, and without X-ray or ultrasound machines my diagnostic capabilities were limited. Because her illness started with a viral respiratory infection, I suspected a secondary infection of some sort. I tried a variety of treatments, and Dorothy seemed to stabilize for a while, but she never returned to normal. She was tired.

For several more months, she enjoyed sitting in the shade at the edge of the forest, especially at "her" corner of the fence line, from where she had an open view in three directions. From her favorite viewpoint she could be sure to see Bouboule, Nama, Jacky, Mado and baby Njabeya, Bikol, and others emerging from their twenty acres of forest along any of a number of chimpanzee trails to join her back at the satellite cage where they would greet her happily, often with hugs. On the opposite side of the fence, she could see the opening of the long trail that led from the center's camp. She was always the first to hear and then see her caregivers bringing food and always the first to know what particular food they were bringing. Through happy grunts and enthusiastic screams she conveyed her excitement, the lesser or greater degree of it depending on how much she liked the food, to her friends who were often out of sight in the forest.

On September 24, 2008, Dorothy lay down on the grass of this favorite spot at the edge of the forest and died. I had left Sanaga-Yong Center for Yaoundé the day before, and Agnes was on vacation in France. Volunteer Monica Szczupider called me at five o'clock in the afternoon to tell me what had happened.

Dorothy came to the cage and ate normally at two o'clock, and only an hour later caregivers Assou and Emmanuel noticed that she wasn't moving in the enclosure. They called Monica, and together they strained to see if Dorothy was breathing. No one thought she was, but they couldn't be sure from their vantage

KINDRED BEINGS

point outside of the enclosure. They couldn't enter the enclosure to approach Dorothy as long as all the territorial males were outside, so they beat the drum to call the chimpanzees inside. The drum normally signals dinnertime, but as it was too early for dinner, the chimpanzees took their time responding. Nama was the first to discover Dorothy, and she was the last to leave her. At one point Monica and the caregivers lost sight of Dorothy as many of the chimpanzees surrounded her. Strong and sensitive Jacky fell on his back, screaming in distress. Bouboule came from the forest, entered the satellite cage, and was closed inside without realizing what had happened. Looking back out and realizing something was wrong, he cried out in fear and confusion. Over the course of more than an hour, all the chimpanzees responded to the persistent drum call. Gradually, one by one, they entered the satellite cage, until only Nama sat beside Dorothy's lifeless body, her hand resting gently on it. Finally, the sight of dinner being served enticed her to sadly enter the cage with the others. When the caregivers got to Dorothy's body, it was already cool.

I raced to get on the 6:00 P.M. train heading back to Sanaga-Yong Center. Agnes caught me on my cell phone in the taxi on the way to the train station. "Hello, Agnes," I answered, recognizing the number showing on the tiny screen.

"Hey, Sheri," she answered back. I knew from the silence following her simple greeting that Monica had called her, too. Agnes, more than anyone else, knew how I felt. We sat in silence for thirty seconds, feeling each other's pain across a thousand miles.

"I'm about to get on the train. I'll call you tomorrow," I told her, and as I hung up slowly, I heard her say, "Okay."

After an eight-hour train ride to the town of Bélabo, followed by a bumpy drive to the Mbargue Forest, I arrived at Sanaga-Yong Center around 3:00 A.M. Dorothy's body was on the floor in the veterinary clinic. I sat beside her on the floor to see her beautiful, peaceful face illuminated by my head torch, to touch her fingers

240

to my face a final time, to feel her coarse hair, and even to smell her body odor, although death had already changed her scent. I knew the heart of this kind chimpanzee—she had earned my admiration and my respect. Alone with Dorothy's body, I cried and said my good-bye. I would never, ever forget her. As much as I didn't want to, I knew at first light I would perform an autopsy. I needed to know why Dorothy died, and there was no one else to do the procedure.

Dawn came, and I put my feelings on hold to do the work at hand. I tried to focus all my attention on liver, intestines, heart, lungs, trying hard to block out the aching fact that they had made up the body of someone I loved. A few minutes into the autopsy, I heard Severin, our guard at the front gate, announce that some-one from the village of Bikol was waiting. I didn't understand the person's name over the crackly radio, and I didn't really care. I guessed it was a sick person, needing medical care. Both Agnes and I administered health care almost daily when we were on-site.

"Not today, for God's sake," I mumbled, although I knew I wouldn't be able to turn a critically ill person away. Through-out the autopsy, I was peripherally aware of Severin's voice on the radio, again and again, announcing the arrival of someone. Deeply sad and tired, I was annoyed at the prospect of having to do any-thing but grieve and bury Dorothy. How the hell many people were waiting for me at the gate, I wondered, and what could be wrong with them that they were coming so early? Were we hav-ing some awful epidemic in the village? When I had finished the autopsy, concluding that Dorothy had probably died of heart dis-ease, and sewed up her body neatly, I reached for the radio with dread. In my American-accented French that didn't hide my fatigue or my impatience, I asked of our gate guard, "Severin, who is waiting at the gate?"

"La population," Severin answered simply. *The population*. It took me all of two seconds to comprehend that people in the community, hearing the afternoon before that Dorothy had died, had walked to our Sanaga-Yong Center camp from their villages miles away, without being invited, to pay respects to her. I crumpled—bowed my head on my arms folded atop the counter—and sobbed again for Dorothy.

The men who had been her caregivers helped me put Dorothy's body in a wheelbarrow, and I covered her with a sheet, leaving only her head exposed. Her facial expression was serene, one of final repose. Raymond conducted a funeral service for the staff, volunteers, and dozens of people from the village community. He spoke of Dorothy's suffering and of the happiness she had known at Sanaga-Yong Chimpanzee Rescue Center. I told the community that the best way to honor Dorothy would be to never eat the meat of chimpanzees again and to speak out against it whenever and wherever they could.

Afterward, we transported Dorothy's body in the creaky wheelbarrow toward her gravesite, which Raymond and the caregivers had prepared beside the grave of Becky. As we stopped at the enclosure to let the congregated chimpanzee friends and family of Dorothy say their final good-byes, Monica snapped her funeral procession photo—that now-famous snapshot of nonhuman grief that created what I hope will be Dorothy's legacy of expanded awareness around the world.

International attention and surprise over the grief of the chimpanzees in that photo inspired me to tell the story of sweet Dorothy and her circle of friends who have impacted my own life to such a large degree. Dorothy was tragically orphaned and then cruelly mistreated for much too large a proportion of her life, but she also knew kindness and love from both humans and chimpanzees. Late in her life, she herself became a mother, a friend, and a kind defender of the abused.

Although Dorothy and I were of different species and we should have lived far apart, each with our own kind in our own habitats, circumstances conspired to introduce us. To an extent, I was included in Dorothy's circle of *people,* and she was included in mine. Thanks to my relationships with Dorothy and the other chimpanzees I have known at Sanaga-Yong Chimpanzee Rescue Center, my acceptance of the deep intelligence and emotional capacity of their species has become such an inexorable part of who I am that it now seems innate. The chimpanzees in the photo include Jacky, Nama, Bouboule, and Mado, each of whom you have glimpsed through my perspective. In these final moments you read from me, I express the hope that you look at the photo a different way now—gone is any surprise over the capacity of chimpanzees to love and grieve, and in its place is a comfortable, albeit perhaps poignant, assumption of normalcy.

EPILOGUE

A year after Dorothy died, her son Bouboule, without political support from anyone in the group, began challenging Jacky for the leadership role. Jacky couldn't beat younger, bigger, and stronger Bouboule in a one-on-one fight, but Jacky had Nama and a small army of adolescents who were willing to fight beside him. Even so, in the face of Bouboule's staunch determination and persistent confrontations, Jacky became increasingly reluctant to engage, and his reluctance was contagious to all except Nama, who never wavered in her support of him. The writing was on the wall for months before it was over, and Agnes and I prayed that Jacky would step aside before he got hurt. Finally, after a year of anxiety on both the chimpanzee and human sides of the fence line, Jacky ended his decade-long reign as alpha male, pant-grunting his submission to his successor.

Unfortunately, Bouboule lacked the political skill to lead well. He dominated through intimidation and fear for about a year, occasionally wounding scapegoats to keep the other males intimidated. Finally, Bikol led the group in a bloody coup against him. The definitive fight occurred late one morning as Agnes and the caregivers stood by the enclosure listening to prolonged screams of battle emanating from the forest. They were helpless to do anything but wait to see the outcome. When the chimpanzees finally emerged from the forest an hour after the fight, Bouboule had multiple wounds, including a severe injury to his testicle, which was bleeding. He was very nervous and obviously afraid of Bikol, who on the other hand was surrounded and supported by many of the females and adolescents in their group. Everyone seemed on edge. With a transparent and compelling desire to make up and be friends, Bouboule made tentative advances toward Bikol, who feigned indifference. Finally, when Bikol allowed Bouboule to close the gap between them and begin grooming him, the nervous tension embodied in all the chimps dissipated. Calm was restored; Bouboule had lost his position to Bikol. The next day I traveled to Sanaga-Yong Center and removed Bouboule's testicle, which was unsalvageable.

Bikol now holds the position of alpha male with a good amount of political support. Arriving in 1999, he was our first baby chimpanzee, and he matured into a kind adult, who has long been a favorite of mine. Unfortunately, he's not an exceptionally strong leader. Although he enjoys his free access to the attractive females, he approaches his responsibilities as authority figure and peacekeeper somewhat halfheartedly. After hearing Agnes disparage his leadership skills for weeks, I was surprised recently when I saw him intervene to prevent bullying twice in one afternoon. That night over dinner when I mentioned it to Agnes, as evidence in defense of my dear Bikol, she said, "Yes, Sheri, he does *something,* but he's not like Jacky." Bikol's name in the local language mean's *the king,* but he seems to be a reluctant one. He enjoys the spoils of power but would rather shun the duties. I hope he'll grow more responsible as he gains in age and experience. He had a good role model in Jacky.

Bouboule has settled down to a relatively passive role in the society and doesn't challenge Bikol. When they hug and comfort one another it reminds me of how they were as young babies. However, there are others who would be king. Moabi and Gabby probably aren't serious contenders, but big adolescents Simon and Future may soon pose threats. Bikol's is a precarious position, and the caregivers make sure he sleeps at night with those who support him.

We lost another piece of the heart and soul of Sanaga-Yong Rescue Center in June 2012 when our beloved Nama, still only in her thirties, passed away after her illness of several months eluded our extensive efforts to diagnose it. Nama's chest kept filling up with fluid, making it difficult for her to breathe. After ruling out heart disease, we finally narrowed the possibilities to cancer and tuberculosis. All the tests we did for tuberculosis were negative, but we still tried treating her for it. I would have tried anything to save her. In the end, nothing worked. At this writing, we're still trying to get autopsy samples out of Cameroon to determine what really caused her illness and death—CITES permits are required to export tissues from endangered species, and they can take months to process. For now, it remains a mystery. My admiration and love for this shining star of a chimpanzee were boundless. I held her in her final moments, and while I celebrate her remarkable life in the telling of her story, I haven't stopped mourning her death that came much too early.

The same week that Nama died, we received badly wounded one-and-a-half-year-old Kanoah, a baby boy who was confiscated during the arrest of a dealer, who is now being prosecuted with the involvement of my friend Ofir Drori's Last Great Ape Organization. A month later, the arrival of eight-month-old baby Carla, brought to us by a Catholic priest, brought our resident chimpanzee population to seventy-three.

Of the first five adult chimpanzees we brought to Sanaga-Yong Chimpanzee Rescue Center, only the elder Jacky survives. Without a lot of responsibility he spends his days playing and relaxing, managing to avoid conflict. He seeks my attention more than before, almost always rising to greet me when I approach, which still flatters and delights me. Unlike before, he solicits me in play, often asking me to run up and down the fence with him. Agnes says jokingly that he has regressed to his lost childhood, and I say he deserves it.

We have begun a series of forest surveys throughout Cameroon to determine if there is a suitable site to reintroduce some of our chimpanzees to a free life. Reintroduction, if it is possible, will be a long and complicated process. It will require a forest site that is good chimpanzee habitat but no longer has chimpanzees living in it, or not many, and at the same time it must not have many humans living around it. Any hunting camps must be removed, and the site must be protectable in the long run, both of which will be the roles of the Cameroon government. Ideally the site should be an important habitat area with conservation value in its own right, quite apart from the reintroduction of chimpanzees. The protection that we would achieve through our reintroduction plan would strengthen ecological diversity or bring it back to a depleted area. Finding such a site is a tall order. We don't know if a site meeting our criteria exists in Cameroon, but to look for it, we've formed a partnership with Ape Action Africa and Limbe Wildlife Center, under the umbrella of Pan African Sanctuary Alliance (PASA), an international organization of which we are all founding members. We are working in collaboration with Cameroon's Ministry of Forestry and Wildlife to conduct forest surveys funded by the Disney Worldwide Conservation Fund.

As for me personally, my worldview and my human relationships have been irrevocably changed by my time with the chimpanzees in Africa. The lens through which I interpret my experiences is much wider now, but at the same time life seems simpler, more basic.

Chimpanzees engage life fully, in the moment. They wear their emotions for all to see, or hear. Even an adult chimpanzee might cry like a baby if he is being rejected, or throw loud and dramatic tantrums over a perceived injustice. A few minutes later, with the proper recognition or comfort, he can be the picture of contentment. The quality of their friendships and family relationships to a large extent determines the quality of their lives. Watching the social vignettes of chimpanzees through the years has taught me to recognize my own pretenses. We are such similar apes. But they bring a primal pureness and immediacy to their expressions of intimacy, which I have come to cherish in my friendships with them. From knowing chimpanzees I have learned to live more honestly and vulnerably.

From my perspective, life is simply too short, even for those who survive to old age, and nothing is guaranteed. None of us knows what breath will be our last. I try to keep that in mind every day—to face each day with interest and a sense of purpose, and to pay more attention to how I affect the momentary experiences of those whose paths I cross. Being an imperfect ape, driven by unseen influences both ancient and contemporary, I cannot say that I'm always kind, but I am almost always compelled toward reconciliation. More than anything these days I'm left with a deep sense of gratitude for this brief but wondrous opportunity to live and love.

ACKNOWLEDGMENTS

Eileen Cope, my compassionate literary agent, tracked me down in Africa to suggest I write my story, nudged me through a year of proposal writing, and then managed to successfully promote an aspiring first-time author. Without Eileen, there would surely be no book. Nancy Hancock, my editor at HarperOne, believed in the book before it was written and applied her considerable talent to make it the best it could be. And I won't forget Elsa Dixon's humane though dogged approach to the editing and titling processes and for her commitment to producing a final product that would make me proud. These people delivered *Kindred Beings,* and I will always be grateful to them.

Agnes Souchal, general manager of Sanaga-Yong Chimpanzee Rescue Center, is both the gentlest and the strongest person I have ever known. I deeply appreciate her consultation, advice, friendship, and encouragement during the writing of *Kindred Beings.*

Karin Cereghino, IDA-Africa program manager, was patiently tolerant of my preoccupation and graciously shouldered more than her share in keeping IDA-Africa afloat with funds coming in during my yearlong obsession with putting the right words, or at least some words, on these pages.

Agnes and Karin as well as Michael Labhard and Malgosia Ceglowski gave generously of their time to read various parts of my early draft and gave very helpful comments.

Edmund Stone and Cindy Scheel, Stan Jones and Cindy Umberger-Jones, Crystal Schneider, Malgosia Ceglowski, and Susan Labhard bestowed me with quiet writing time, and my daughter, Annarose, with hours of enjoyment in their company.

Photographs for the book were contributed by a talented array of photographers including Carol Yarrow, Mirjam Schot, Monica Szczupider, Agnes Souchal, Marie-Eve Lavigne, Leslie Kadane, Jacques Gillon, and Karl Ammann. I especially thank Monica for the extra efforts that landed her photograph of Dorothy's funeral in *National Geographic*.

Annarose Sara Muna, my kindhearted, independent-minded eleven-year-old, cheered me on each day, cooked and ate countless breakfasts alone, and spent many weekends quietly entertaining herself.

In addition to those who provided direct support for *Kindred Beings*, there were many people who made possible my work for chimpanzees and the life-changing experiences that were its subjects.

When the chimpanzee sanctuary in Cameroon was only a vision in my head, Dr. Elliot Katz, president emeritus of In Defense of Animals (IDA) International, had faith in me. IDA-Africa and Sanaga-Yong Chimpanzee Rescue Center are included in his legacy.

Edmund Stone cofounded IDA-Africa as a program of IDA International and the U.S. base of support for Sanaga-Yong Chimpanzee Rescue Center. He worked tirelessly for over a decade to raise funds for the work in Cameroon. Edmund's wife, Cindy

Scheel, has been both a valued development consultant and a fabulous caterer for our fund-raising events.

Estelle Raballand brought her boundless energy, courage, and bilingualism to our early efforts in Cameroon and taught me much about chimpanzees.

George Muna is the most genuinely generous person I have ever known and his contributions to my endeavors for chimpanzees in Cameroon are countless. I thank him for all of these and most of all for being a loving parent to our daughter, Annarose.

Karen Bachelder deserves special mention for traveling to Cameroon to volunteer at Sanaga-Yong Chimpanzee Rescue Center eleven times, for generously managing the volunteer selection process, and for her sage and well-considered advice, from which I have benefited on many occasions.

When they were directors of Ape Action Africa, the late Colonel Avi Sivan and Talila Sivan contributed immeasurably to my safety, security, and peace of mind in Cameroon.

Rachel Hogan and Babila Tafon, current director and manager of Ape Action Africa, have been frequent collaborators and have assisted Agnes and me many times on issues of security, chimpanzee transportation, and medical care.

It has been my pleasure to work with still other truly unique, larger-than-life people whose collaboration at various times in the past fifteen years facilitated and advanced my work for chimpanzees in Cameroon. I owe debts of gratitude to Shirley McGreal, Peter Jenkins, Liza Gadsby, Ofir Drori, Dave Lucas, Karl Ammann, and Jean Liboz.

Many valued colleagues have contributed to my work through Pan African Sanctuary Alliance (PASA). They include Doug Cress, Anne Warner, Julie Sherman, Steve Unwin, Christelle Colin, Kay Farmer, Ainare Idoiaga, John Kyang, Felix Lankester, and all the other managers and directors of the PASA sanctuaries in Africa.

ACKNOWLEDGMENTS

The employees of Sanaga-Yong Chimpanzee Rescue Center have been essential since 1999, and their number has steadily increased. I single out for mention those who have been with us a long time and/or forged a special bond with the chimpanzees: Raymond Jules Tchimisso Guea (a member of our management team), Timothy Maishu Wirba, Emmanuel Ndong Mene, Assou Felix Francois, Julien Clerck Gomdong, Bertrand Avom, Barnabe Julien Andang, Henriette Nganyet, Nicholas Banadzem, and Severin Bipan.

Thousands of generous and caring people in the United States and other countries around the world have contributed financially to the work we are carrying out in Cameroon, and though they number too many to mention by name, from my heart I appreciate them all. There are some who, in addition to financial support, have given very generously of their time to help us raise essential funds. They include Brian Behrens, Steven Bernheim, Marianna Boros, Malgosia Ceglowski, Christine Desvignes, Claudine Erlandson, Ruth Fredine, Julia Gallucci, Al Hainisch, Betsy Holbrook, Julie Honse, Sangumithra Iyer, Kerri Jackson, Mohamed Jantan, Stanley Jones, Cindy Umberger-Jones, Leslie Kadane and Kyle Doane, Erika Knauf Santos, Andrea Kozil, Susan Labhard, Jessica Martinson, Molly Mayo, Laura Michalek, Iain Moffat, Heather Murch, Perrine Odier, Mary Perin, Rebecca Pool, Meg and Jon Ratner, Gwendy Reyes-Illg, Suzanne Roy, Richard Satnick, Crystal Schneider, Valerie Sicignano, Franz Spielvogel (Laughing Planet), Chuk and Donna Steadman, Connie Theil, Dana Vion, Kimber Webb, Rachel Weil, and Jacque West.

CREDITS

Grateful acknowledgment is given to the following photographers for the use of their work in this publication:

Ann de Graef: p. 12 *(top)*
Jacques Gillon: p. 7 *(top)*
Marie-Eve Lavigne: p. 1
Agnes Souchal: pp. 2, 6 *(bottom)*, 7 *(bottom)*, 9 *(top and bottom)*, 13, 14 *(top)*
Sheri Speede: pp. 5 *(bottom)*, 8, 10, 11 *(top)*
Monica Szczupider: pp. 9 *(middle)*, 11 *(bottom)*, 14 *(bottom)*, 15
Carol Yarrow: p. 16

INDEX